Genes: A Very Short Introduction

VERY SHORT INTRODUCTIONS are for anyone wanting a stimulating and accessible way in to a new subject. They are written by experts, and have been translated into more than 40 different languages.

The Series began in 1995, and now covers a wide variety of topics in every discipline. The VSI library now contains over 350 volumes—a Very Short Introduction to everything from Psychology and Philosophy of Science to American History and Relativity—and continues to grow in every subject area.

Very Short Introductions available now:

Available soon:

For more information visit our website

www.oup.com/vsi/

Jonathan Slack

GENES

A Very Short Introduction

OXFORD
UNIVERSITY PRESS

OXFORD

UNIVERSITY PRESS

Great Clarendon Street, Oxford, OX2 6DP,
United Kingdom

Oxford University Press is a department of the University of Oxford.
It furthers the University's objective of excellence in research, scholarship,
and education by publishing worldwide. Oxford is a registered trade mark of
Oxford University Press in the UK and in certain other countries

First edition published in 2014

Impression: 1

Published in the United States of America by Oxford University Press
198 Madison Avenue, New York, NY 10016, United States of America

British Library Cataloguing in Publication Data

Data available

Library of Congress Control Number: 2014940246

ISBN 978-0-19-967650-7

Printed in Great Britain by
Ashford Colour Press Ltd, Gosport, Hampshire

Contents

Preface

What is a gene? In essence it is a molecule of DNA, present in every one of our cells, that controls the synthesis of one particular protein in our bodies. But this simple definition does not do justice to the richness of the gene concept and its many ramifications throughout the life sciences. For example, some genes do not code for proteins, some have no function at all, others are presumed to exist but have not been identified in terms of DNA.

Everyone has heard of genes and we have come to feel that they are fundamental to who we are. But there is also a lot of uncertainty and confusion about what this means. For example, does having cancer in a first degree relative mean I have 'cancer genes'? If I have genes also found in Stone Age fossils, does this mean I am very primitive? Does the existence of 'selfish genes' mean that human nature is inherently selfish?

We also know that genes have become the basis of a huge technology which generates pharmaceuticals, diagnostic tests, forensic identification, paternity tests, and GM crops, which are welcomed by some people and feared by others. Many are suspicious of the gene concept as applied to race, intelligence, criminality, or other human attributes, while others presume that these things are mostly under genetic control most of the time. Many have a desire to understand this entity which penetrates so deeply into their lives but still seems rather mysterious and confusing.

This book is not a textbook of genetics, but it is a brief introduction to the various conceptions of the gene currently used in the life sciences. The aim is to enable readers to appreciate the main ideas about genes, evaluate contentious issues, and to navigate to more advanced texts if they wish.

List of illustrations

Chapter 1
Genes before 1944

In 1938 a remarkable pair of articles was published in the *Quarterly Review of Biology*. They were by an American biology professor from the University of Missouri, Addison Gulick, and they were about the nature of the gene. These articles are rarely consulted today since they were written shortly before it was discovered that genes consisted of deoxyribonucleic acid (DNA). However, they are remarkable in showing how much was known about genes even before their chemical nature was established. Gulick knew that genes were located in the chromosomes of the cell nucleus, and were complex structures that somehow directed the synthesis of enzymes and the development of the organism. He knew that they normally remained stable from one generation to the next, and that occasional changes, called mutations, could spread through the population and be the basis of evolution by natural selection. He also made surprisingly accurate estimates of the sizes and numbers of genes in various types of organism. These articles illustrate that to appreciate how our current understanding of the gene came about we need to go back much further than the famous 'double helix', discovered by Watson and Crick in 1953.

Two completely separate lines of work led to our modern view, and they came together shortly after the appearance of Addison Gulick's articles to create the new science of molecular biology.

One was the study of heredity by biological experimentation and the other was the study of the chemistry of DNA.

Biology of heredity

Before the 18th century there was little informed speculation about heredity. Even the word did not exist ('heredité' first appeared in France, 'genetic' in England, both about 1830). Before then there was plenty of animal breeding and vague ideas of 'blood lines', but this was uninformed by much understanding of reproduction. In the 18th century the first systematic breeding of agricultural animals had begun. Robert Bakewell, a sheep breeder from Dishley, near Loughborough, England, bred a line of New Leicester sheep that grew faster and produced more meat than before (Figure 1). This was done by mating the best males and females to create a self-reproducing population (breed) that

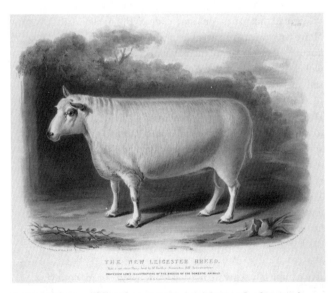

THE NEW LEICESTER BREED.

1. New Leicester sheep. From David Low's *The Breeds of Domestic Animals of the British Islands*, London, 1842

maintained the new characteristics in a stable way. The experience of animal breeding conveyed the idea that heredity involved the blending, or averaging, of the distinct characters, also known as traits, of the parents. It was quite understandable that animal breeders should have believed in the blending of traits since this is what you see when animals are mated and characters such as height or weight or growth rate are measured. But the blending theory of inheritance was to become a serious problem for Darwin's theory of natural selection.

By the time of Darwin's work in the mid-19th century, the fact that biological evolution had occurred was reasonably well accepted by scientists, mostly on the basis of the changes seen in the fossil record. The real impact of Darwin's work, and that of his contemporary Alfred Wallace, was to provide an actual and credible mechanism for the changes seen in living organisms over evolutionary time. This mechanism was natural selection, and the case for it is simply stated. If a population of animals or plants varies with regard to some traits, if those traits are heritable, and if they affect the likelihood of reproduction, then the composition of the population will inevitably shift between each generation. The traits associated with more reproduction will become more common, and will eventually displace the alternatives. The direction and speed of the shift will be determined by the selective conditions that cause the differential reproduction of the individuals with the different traits. The theory of natural selection seems very compelling, and it is especially compelling as presented by Darwin in *The Origin of Species* (1859), which contains a huge array of examples drawn from natural history to support the case. At the time, the main opposition to the theory was from religious groups who realized that the principle of natural selection undermined the 'argument from design', an important argument for the existence of God, and also from those offended by the idea of a biological kinship between humans and animals. However, there was also some scientific opposition, and the most serious was that which focused on the difficulty of reconciling natural selection with blending inheritance.

Supposing that one individual is slightly better suited to reproduction than others due to the possession of a particular trait. Because the favourable trait is rare, he or she will most likely mate with an individual lacking it and their offspring will then have it in a diluted form. After three or four generations the hereditary factors responsible for the trait will be diluted out almost completely. So selection has only a few generations to operate and this will not be enough to change the whole breeding population unless the reproductive advantage conferred by the new trait is very large indeed. Darwin himself was well aware of the problem but he was also opposed to the idea of large jumps in evolution and favoured the idea that evolution progressed in a smooth and imperceptible manner via many small changes.

Some thinkers followed this argument to its logical conclusion and concluded that the hereditary factors responsible for evolution must have large effects, such that substantial selection could occur before they became diluted out. In this way the trait might become common enough for some matings to occur between parents who both possessed it and it would no longer be diluted in their offspring. Among these thinkers was William Bateson who collected in his book *Materials for the Study of Variation* (1894) a remarkable set of examples of discontinuous and qualitative variation within animal and plant populations. More direct evidence for the existence of large heritable changes came from observation of spontaneously occurring 'mutations'. In particular the Dutch botanist Hugo de Vries in 1886 observed the *de novo* appearance of dramatically new forms of the Evening Primrose, which bred true in subsequent generations. Nonetheless, blending inheritance remained a serious problem for the theory of natural selection.

In fact the solution to the problem had been provided as early as 1866 by Gregor Mendel, a monk at the Abbey of St Thomas in

Brno, now in the Czech Republic. In the early 19th century, Brno was a centre of textile manufacture and of sheep breeding, and the Abbey already had a two-hectare experimental garden. Mendel had received education about animal and plant breeding in the course of his studies of philosophy at the University of Olomouc, and he was encouraged to continue his work at Brno by the Abbot. Between 1856 and 1863, Mendel conducted a number of experiments with peas. He was fortunate enough to choose simple characters, which we would now call characters determined by single Mendelian genes, rather than complex characters determined by many genes. Among these characters were a round or wrinkled appearance, and a green or yellow colour. Mendel postulated that there were invisible hereditary 'factors' causing each visible character, and showed that there were predictable rules for their inheritance. His breeding experiments indicated that each individual plant contained two factors for each character, one derived from each parent. When reproductive cells (pollen or eggs) are formed, each contains just one factor, randomly selected from the two possibilities available in that plant. In some cases one factor would suppress the other: we should now call this a dominant gene. For example a cross between yellow and green peas gives only yellow offspring. However if these offspring are crossed to each other, then 25 per cent of the next generation are green, indicating that the green factors are still there, but cannot be expressed in the presence of the yellow factors (Figure 2). So, Mendel showed that the hereditary factors behaved as discrete units, such that each parent provided one to each offspring and the appearance of the offspring depended on the specific combination of factors inherited and the dominance rules between them. Mendel published his work in the *Verhandlungen des naturforschenden den Vereines in Brünn* in 1866. But this was what we should now call a 'low-impact journal', and nobody noticed. After he became Abbot in 1868 he was mostly occupied with administrative duties. He died in 1884 with the wider world still being ignorant of the founding principles of genetics that he had discovered.

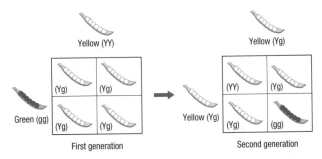

Yellow (YY)

Yellow (Yg)

(Yg) (Yg)

(Yg) (Yg)

Green (gg)

Yellow (Yg)

First generation

(YY) (Yg)

(Yg) (gg)

Yellow (Yg)

Second generation

2. Mendel's peas. When yellow peas are crossed to green peas, the first generation seeds are all yellow. But when members of the first generation are crossed to each other, 25 per cent of the second generation seeds are green. Mendel explained this by postulating factors, here called Y for yellow and g for green, such that the yellow factor is dominant over the green where they occur together

The 20th century

In Western Europe and America, arguments continued over whether natural selection could work through blending inheritance, and whether mutations of large effect were a credible source of variation. Not until 1900 was Mendel's work rediscovered and further developed. Several people were responsible for this including Hugo de Vries of mutation fame and the German botanist Carl Correns. It was immediately apparent that the major contradiction had now been removed. Mendel's factors were stable and persisted from generation to generation and variation in a population existed because of the differences between the factors that were present in each individual. So it was not necessary to postulate new mutations to explain every newly appearing variation. Moreover the more complex characters, whose inheritance appeared to be of a blending nature, could be explained as resulting from the action of several independent Mendelian factors.

In the late 19th century improved microscopes and new stains from the chemical industry had improved visualization of cells and their nuclei. The German anatomist Walther Flemming, working at Kiel, first identified chromosomes, and described the process of cell division, now called mitosis, in which the chromosomes enter the nuclei of both daughter cells. The Belgian cytologist Edouard van Beneden showed that there was a characteristic chromosome number for each species and that this number was found in the various different cell types of an organism. He followed chromosomes of the nematode *Ascaris* through cell division and showed that the number was conserved, but that it halved during the formation of reproductive cells (divisions forming reproductive cells are now called 'meioses'; singular, 'meiosis').

By the beginning of the 20th century, after Mendel's work had been resurrected, Theodor Boveri in Germany and Walter Sutton in the USA independently showed that chromosomes behaved just like Mendel's hereditary factors. From then on most scientists believed that the chromosomes were the hereditary factors or at least contained them. The hereditary factors themselves were named 'Gene' (in German this is plural, equivalent to 'genes' in English) in 1909 by Wilhelm Johannsen, Professor of Plant Physiology at the University of Copenhagen. The term was derived from the Greek γενεα (=generation or race). This is an interesting linguistic example of the entity being named after the process, as 'genetic' had been in use since 1830, and 'genetics', as a noun, was introduced by William Bateson in 1905.

The centre of gravity then moved to the USA where Thomas Hunt Morgan, at Columbia University, effectively established modern genetics through his studies of the fruit fly *Drosophila*. *Drosophila* is a small insect with a short generation time. Large numbers can be kept in small tubes, many crosses can be carried out in a reasonable time scale, and insects have a lot of complex anatomy on the surface that it easy to observe by simple visual examination. *Drosophila* is therefore very well suited to genetic work in the laboratory. Despite much effort, no treatments were found that induced new mutations, but a number of spontaneous mutations were discovered and used for breeding studies. Whereas Mendel's factors had separated from one another in terms of their representation in individuals of the next generation, most of Morgan's *Drosophila* mutations did not do so. The difference is that *Drosophila* has only four chromosomes, so it is quite likely that two genes will lie on the same chromosome and thus be transmitted together to reproductive cells. However genes on the same chromosome do sometimes separate and Morgan was able to work out that this occurred because the two parental chromosomes always form pairs with each other in the last cell division (meiosis) leading to the formation of reproductive cells (pollen, sperm, eggs). During this phase of pairing the chromosomes can break and rejoin such that segments are exchanged. So a gene that was on one parental chromosome can join one that was on the other, and be transmitted together into the reproductive cells. Importantly the probability of separation of two genes depends on the distance between them. So the frequency of segregation of genes on the same chromosome became the basis for genetic mapping. Using these techniques Morgan's group created accurate maps of the *Drosophila* chromosomes showing the positions of many genes for which visible mutations were available.

One of Morgan's students, Hermann Muller, by then at the University of Texas, finally discovered a treatment for *Drosophila* that would generate new mutations. This was X-irradiation. After irradiation the chromosomes often showed visible changes, strengthening the link between chromosomes and genes. In the early 1930s it was found that the salivary glands of the *Drosophila* larva contained giant chromosomes, much larger than those found in most other cell types. These showed vastly more detail down the microscope than normal-size chromosomes and when suitably stained they were seen to contain thousands of bands, which were suggestive of being single genes or small groups of genes. As a result of this programme of work on *Drosophila*, Morgan's school had definitively proved that genes lay on chromosomes, that they were arranged in a linear manner, and that each cell contained a chromosome set from each parent such that there were two copies of each in cells of the body and one copy of each in reproductive cells.

In the late 19th century another scientific tradition had grown up which we should now call statistics. In relation to evolution, a school surrounding Karl Pearson (the 'biometricians') held to the Darwinian orthodoxy and considered that inheritance was inherently quantitative and continuous and that evolution proceeded in small steps. Although apparently at variance with Mendelism, the two traditions were eventually reconciled by the work of R.A. Fisher. Fisher was for many years director of the agricultural experimental station at Rothamsted, England, where he devised most of the statistical methods now used for designing and analysing quantitative experiments. He is also the father of modern quantitative genetics. Fisher showed that quantitative characters, such as height and weight, could be accounted for by the action of several Mendelian genes, each with variants of different effect, and that that the blending behaviour seen in such cases was completely consistent with Mendelian genetics. With Sewall Wright, of the University of Chicago, and J.B.S. Haldane, then at Cambridge University, Fisher created what had since been known as 'the modern synthesis', which is essentially the theory of natural selection in terms of Mendelian genetics.

During the early 20th century there was great parallel increase of understanding about how living organisms transformed foodstuffs into energy and into the materials for growth. It had been shown that each metabolic step was one chemical transformation catalysed by an enzyme and that enzymes themselves were proteins. George Beadle and Edward Tatum, working at Stanford University, chose the bread mould *Neurospora crassa* as more amenable than *Drosophila* for biochemical studies. *Neurospora* can be grown on simple agar-based media and if a mutation causes the inability to synthesize a particular substance, the mould can still grow so long as this substance is added to the medium. Beadle and Tatum created many mutant strains by X-irradiation and showed that each mutation led to the lack of a specific enzyme. They concluded that genes either were enzymes, or controlled their formation. This was later encapsulated in the slogan 'one gene, one enzyme'.

So, by the beginning of the Second World War a lot was known about genes. It was understood that they lay on chromosomes in the cell nucleus, that each chromosome came as a pair, with one member from each parent, and that each gene somehow was responsible for the production or activity of a specific enzyme. It was known that changes in genes, mutations, could be evoked by X-rays, and also by certain chemical substances which were therefore called mutagens. What was not known was the chemical nature of the gene. What were genes actually made of? When Addison Gulick wrote his reviews this was a major issue. Like most others at the time he presumed that they were large molecules, probably consisting of proteins or protein complexes. But the identity of the molecules, how they could be replicated, and how they could control the formation of enzymes, remained a complete mystery.

DNA

Contrary to general belief, DNA has been known to biochemistry for a long time. It was discovered in 1869 by Friedrich Miescher, a Swiss physiological chemist working at the University of Tübingen in Germany. Miescher studied white blood cells which he obtained in large quantities by squeezing the pus out of the bandages of surgical patients. He isolated nuclei from the cells using the enzyme pepsin, then extracted them with weak alkali and precipitated with acid. The resultant substance, called nuclein, was found to contain phosphorus, but was quite unlike the other biological substances known at the time to contain phosphorus. Miescher considered the function of nuclein to be simply a store for phosphorus within the body. Subsequent progress was quite slow by modern standards. In contrast to biology, where a breeding experiment could be conducted almost as easily in 1880 as 1980, in chemistry the techniques of the late 19th century were very primitive. There were few methods for separating substances, with precipitation using salts or solvents being generally employed. There were also few methods for analysing even simple molecular structures, the mainstay being the purification of the substance followed by a determination of the proportions by weight of each chemical element that it contained. The difficulties were particularly severe in biological chemistry which deals with large molecules that are easily damaged by harsh conditions.

In the period 1880–1900, Albrecht Kossel working at the University of Berlin conducted extensive studies on nuclein. During this time the substance became known as 'nucleic acid' because of the acidity arising from its high content of phosphate groups. Kossel found that major constituents of nucleic acids were the five bases: adenine, thymine, cytosine, guanosine, and uracil. In chemistry a 'base' is an alkaline substance, and all of these substances are weakly alkaline because of the amino (NH_2) groups that they contain. They are usually represented by the single letter abbreviations: A, T, C, G, U. In the early 20th century extensive further chemical studies were made by Phoebus Levene at the Rockefeller Institute for Medical Research in New York. He identified the sugar ribose, and later its close relative deoxyribose, as components of nucleic acids. Most importantly, he worked out that nucleic acids were long molecules in which the bases, sugars, and phosphates were joined together in a chain. By 1935 the actual chemical structure of nucleic acids was essentially finalized with the identification of the positions of the base, and the attachments of the two phosphates on each sugar molecule (Figure 3). Without going into details, it will be appreciated that the potential number of ways of joining bases, sugars, and

3. **DNA 2D structure. This shows a chemical structure of five nucleotides only. Each nucleotide consists of the sugar which contains a ring of four carbon and one oxygen atoms, attached to a base, depicted here just as A, C, G, T, and to the adjacent sugar by a phosphate group. 5′ and 3′ are numbers given to the different carbon atoms of the sugar**

phosphates together is huge, and the discovery that bases lay on the 1′ position of the sugar and phosphates on the 3′ and 5′ positions (the primed numbers indicating the individual carbon atoms of the sugar), was a major advance in understanding the structure.

In terms of chemical nomenclature, a *nucleoside* consists of a base joined to a sugar, a *nucleotide* consists of a nucleoside joined to a phosphate group, and a *nucleic acid* consists of several nucleotides joined together, as shown in Figure 3. During the 1920s it had become appreciated that there were two major classes of nucleic acid. 'Animal nucleic acid' (now called deoxyribonucleic acid or DNA) contained deoxyribose as the sugar and A, C, G, and T as bases. 'Plant nucleic acid' (now called ribonucleic acid or RNA) contained ribose as the sugar and A, C, G, and U as the bases. In reality animals and plants both contain DNA and RNA, but DNA is found mostly in cell nuclei and RNA in the surrounding cytoplasm. The names arose because DNA had been generally extracted from calf thymus which contains a lot of cells with large nuclei and little cytoplasm, while RNA was usually extracted from yeast which has a small nucleus and a lot of cytoplasm. By the end of the 1930s it was understood that the 'animal' and 'plant' nomenclature was not appropriate.

So, by the time of Addison Gulick's reviews, a very accurate picture of the gene had been built up from biological studies, and the chemistry of DNA and RNA had been largely worked out. So why did nobody realize that genes were made of DNA? There were in fact two very good reasons. First, the most popular view about the structure of DNA was that it was a tetranucleotide, in other words each molecule consisted of one of each of the nucleotides containing A, C, G, and T. This was believed because when DNA was degraded, approximately equal amounts of each of the four bases were recovered. Also measurements of the molecular weight of DNA indicated a value of about 1,500, equivalent in size to four nucleotides joined together. A tetranucleotide would be a rather simple structure compared to the proteins which were known to be much larger and more complex molecules, and it seemed impossible for genetic information to be stored in something as simple as a tetranucleotide. On the biological side, chromosomes were known to contain genes and they also contained DNA. However, when cells in the course of division were viewed down the microscope using stains that reacted specifically with DNA, the colour was lost during the cell division process. This suggested that the DNA became degraded and resynthesized in every cell cycle, hardly behaviour compatible with the known long-term stability of the genes.

The chemical nature of the gene

Some of the most exciting discoveries in the life sciences have been of the type where a particular substance is identified as being responsible for a biological phenomenon. In this category belong hormones, vitamins, and enzymes. But perhaps none was quite so dramatic as the discovery of the chemical nature of genes. Genes are, after all, at the absolute centre of biology. Viewed with hindsight, discovery may seem painfully slow. But at the time, discovery of anything is remarkably difficult, and genes were no exception. After all, there was no guarantee that any one class of substance did make up the genes. Perhaps genes were made of many complex substances and it was the organization of huge molecular complexes that was all important. Perhaps even if pure genes could be analysed in terms of their constituent molecules there would be no clue as to how they actually worked.

In any case, you cannot identify a substance without a means of measuring its activity. In biochemistry this means you need a suitable bioassay, a procedure in which biological activity can be demonstrated and measured. In the 1930s there seemed no obvious way in which a gene might be put into a *Drosophila* or a *Neurospora* in order to change the properties of the organism. The bioassay for genes would need to involve something much simpler.

It had actually already been invented by Frederick Griffith working at the British Ministry of Health Pathological Laboratory. Griffith was studying the bacterium *Streptococcus pneumoniae*, a major cause of human pneumonia as well as many other infections. These bacteria come in various subtypes. The 'smooth' forms have slimy capsules around the cells that enable them to evade the immune system of the host, while the 'rough' forms do not. Accordingly the smooth forms are much more pathogenic. The names 'smooth' and 'rough' actually relate to the appearance of colonies of the bacteria when grown on agar plates. The key experiment of Griffith, published in 1928, was to show that when mice were infected with rough bacteria plus heat-killed smooth ones, they would die, and then live, true-breeding, smooth bacteria could be recovered from the carcass. This did not occur with rough bacteria alone, or with heat-killed smooth bacteria alone. The implication was that genes responsible for the formation of the smooth capsules had passed from the dead bacteria to the live ones.

The work was taken further by the lab of Oswald Avery and various coworkers at the Rockefeller Institute for Medical Research, New York. An important step forward was to show that the effect would work in vitro. Rough bacteria treated with chemical extracts of the smooth type would indeed yield smooth colonies on agar plates, so the phenomenon did not require intact mice with all the complexities of infection and immunity that are going on in a whole animal. Eventually the extracts from the smooth bacteria were purified to a high degree and were shown to conform to the chemical composition of DNA. Moreover the activity could be destroyed by specific DNA-degrading enzymes, but not by other treatments. The seminal paper, published in 1944, is very cautious in its interpretation, and it only speculates about the identity of the 'transforming principle' with a gene. It did not create any great sensation at the time and was only moderately cited in other works. However, its conclusions had been noted by two individuals who were to play a big role in events to come. They were Francis Crick and James Watson, and Avery's paper convinced them that DNA was really, really, important stuff.

Chapter 2
Genes as DNA

The double helix

After 1944, and as a result of Avery's experiments, it was known to a small group of cognoscenti that at least one type of gene, found in the *Pneumococcus*, consisted of deoxyribonucleic acid (DNA). Over the next 20 years a remarkable set of discoveries established the overall shape of modern molecular biology and this story has been core textbook material ever since. Best known among these events was the discovery of the three dimensional structure of DNA: the famous double helix.

The chemical studies of the previous hundred years had shown that the structure of the DNA molecule consisted of a long chain of the four nucleotides, A, T, C, and G, joined by phosphate groups. But there was an important aspect of the chemical structure that was still unknown: the three-dimensional arrangement of these components in space. This was deduced by a combination of X-ray diffraction studies carried out by Maurice Wilkins and Rosalind Franklin of King's College London, and molecular model building by James Watson and Francis Crick at the University of Cambridge. Their work was recognized by the award of the Nobel Prize for Physiology/Medicine in 1962 (Rosalind Franklin was not a prize winner due to her early death in 1958, but her achievement is considered seminal).

The model building work was greatly helped by a critical piece of chemical information that had recently been discovered by Erwin Chargaff at Columbia University. This was the fact that DNA contained equal numbers of molecules of A and T, and equal numbers of C and G. On the other hand, the ratio of A+T/C+G varied with the organism from which the DNA was isolated. Previously it had been thought that all four nucleotides occurred in equal proportions, hence the tetranucleotide hypothesis, but Chargaff's measurements were more accurate than those of the 1920s and 1930s and suggested that A molecules must be in some way associated with Ts, and Cs associated with Gs.

Although molecular structure does not always explain function, it turned out that the three dimensional structure of DNA did, in fact, explain how the substance could act as the genetic material. One molecule of DNA consists of two strands, lying in opposite senses in a helical configuration. They are held together by attractive forces (hydrogen bonds) between the bases, such that A pairs with T and C pairs with G (Figure 4). This base pairing was consistent with 'Chargaff's rule': A–T and C–G. Because the structure consisted of two complementary strands this immediately suggested that DNA could be replicated by separating the two strands and assembling a new complementary strand on each, such that a new A is inserted opposite every T and a new C opposite every G. The drama of the events leading up to the discovery of the structure of DNA is nicely recounted in James Watson's book *The Double Helix* (1968).

Evidence that DNA actually did replicate in the way proposed was soon obtained. Prior to each cell division, the entire DNA content of each chromosome becomes copied into two double helical molecules, each containing one of the original strands and one newly synthesized strand. When cells divide, one of the two identical double stranded molecules goes to one of the daughter cells and the other copy goes to the other daughter. Hence both daughter cells have exactly the same genes as the original cell.

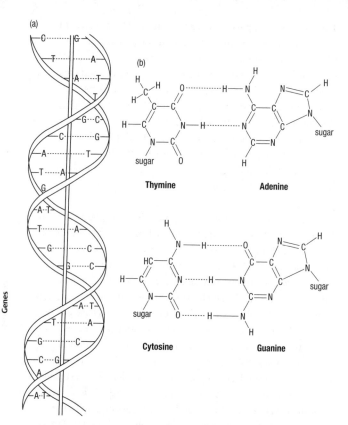

4. DNA 3D structure. (a) DNA is a double helix consisting of two strands running in opposite directions. The strands are held together by hydrogen bonds between the bases; (b) Base pairs formed by A–T and C–G

It was known that genes were responsible for making enzymes and other proteins. Proteins consist of chains composed of amino acids of which there are 20 main types. How could a sequence of four types of unit: A, T, C, G control formation of another sequence drawn from 20 types of unit? This became known as the problem of the genetic code. DNA has four different kinds of base so there are 16 possible permutations of two bases and 64 possible permutations of three bases. From this calculation it was early reasoned that it would need at least three bases to specify a particular one of the amino acids. The actual chemical machinery for making protein in accordance with the DNA sequence turned out to be very complex and to involve several kinds of ribonucleic acid (RNA) (Figure 5). First the gene, which is a sequence of nucleotides on one strand of the DNA double helix, is copied into messenger RNA in a process called *transcription*, which is carried out by a number of enzymes and other proteins. During

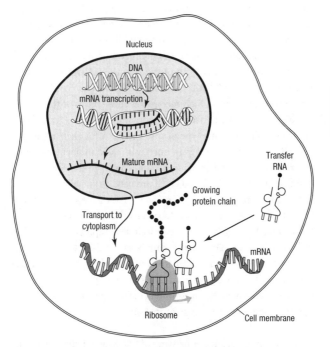

5. Protein synthesis. First the sequence of the gene is copied into messenger RNA (mRNA). This moves to the cell cytoplasm and a protein chain is assembled according to the sequence with one amino acid represented by each of the triplets of 3 nucleotides

transcription the RNA strand is assembled by placing each of the
four bases opposite its complement in the DNA sequence. RNA is
a single stranded molecule and also has four bases except that
uracil (U) is used in place of the thymine (T) in DNA. During
transcription U is inserted opposite A in DNA, and the other
pairs are as usual: A goes opposite T, C opposite G, G opposite C.
The messenger RNA then moves from the cell nucleus, where the
DNA is located, into the cytoplasm, where it associates with
ribosomes. These are complex bodies each consisting of three
or four molecules of RNA together with about one hundred
specific molecules of protein. The RNA of ribosomes does not
have an informational function like messenger RNA; it is
structural, serving to hold the ribosome together and to carry out
its catalytic functions. In order to assemble a protein, a number
of amino acid molecules need to be joined into a linear chain.
First, individual amino acid molecules in the cytoplasm become
attached to transfer RNAs. The transfer RNAs each contain
a three base sequence which recognizes three bases of the
messenger RNA by the usual complementarity rules (A binds U;
C binds G). There is one type of transfer RNA for each triplet.
Each tRNA molecule brings in one amino acid molecule which
becomes joined to the growing protein chain. Of the 64
permutations of A, C, U, and G, 61 are used to specify amino
acids, and are recognized by specific transfer RNAs. The other
three are used as terminators, to specify the end of the protein
chain. Once this is reached, the protein is complete and detaches
from the ribosome. The relationship between the 64 base triplets
in messenger RNA and the 20 amino acids is known as the *genetic
code*. The code is 'degenerate' in that several triplets can specify
the same amino acid. Some amino acids have as many as six
triplets, others have four, two, or just one.

The intellectual package consisting of the double helical structure
of DNA, its semiconservative replication, messenger RNA, and the
genetic code, comprises the intellectual foundation of modern
molecular biology. These facts were discovered by a relatively

small group of people in the ten years following the discovery of the double helix. Since the 1960s the structure of genes and the mechanisms of gene activity and protein synthesis have been studied in enormous detail by hundreds of thousands of people. The current textbook picture is very complex and involves thousands of individual molecular components. However the essence of the process remains as described here.

The second generation of molecular biology involved its conversion into a practical toolkit for analysing and modifying living organisms. This occurred later, in the 1970s and 1980s, and involved the introduction of a number of techniques that have now become extremely powerful. They include molecular cloning, whereby single genes can be isolated and expanded to a quantity suitable for practical applications. Genes can be introduced into organisms by a whole range of methods, often based on the use of viruses that normally introduce genes into bacteria or animal cells. Also in the molecular biology toolkit is DNA sequencing, which is a suite of chemical techniques for determining the order, or 'sequence', of A, T, C, and G in the molecules of a DNA sample. Sequencing techniques have become automated and massively powerful such that, at the time of writing, the sequence of a whole human genome can now be determined in a few hours for a few thousand dollars. In addition there are several sensitive techniques for measuring the amounts of specific genes or gene products in cells and tissue samples. These depend on the recognition of complementary nucleic acid molecules (hybridization) and on amplification of samples by a method resembling normal DNA replication (polymerase chain reaction). Such methods enable the accurate measurement of very small quantities of specific types of DNA or RNA in biological samples.

What is a gene?

This is a question often set as an essay topic for starting undergraduates in the life sciences. In one sense the answer is simple: 'a gene is a piece of DNA'. But even within the framework of molecular biology things are much more complex than this. A sequence within the DNA that actually specifies a protein in the way described above is called a 'coding region'. Even the relatively simple genes of bacteria possess some regulatory sequences in addition to the coding region, and genes in animals or plants are much more complex. Only a small fraction of their total DNA consists of protein-coding genes and only a small part of these genes are actually coding regions. In animals and plants, genes are split up into several segments called exons. In between the exons are substantial lengths of sequence called introns. The RNA transcript starts somewhat before the triplet for the first amino acid and finishes somewhat after the termination triplet, including both exons and introns. The introns are then spliced out to generate a shorter mature messenger RNA, containing only the sequences complementary to the exons. This mature messenger RNA is transported to the cell cytoplasm. Even in the mature messenger RNA there is a region at each end that does not code for protein. 'The gene' (Figure 6) is usually considered to include the introns, the promoter sequence where transcription begins, and sometimes also the regulatory sequences in the DNA, controlling gene expression. The regulatory regions which determine in which cell types the gene is active can be quite complex, and much larger than the coding region itself.

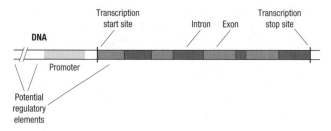

6. Gene structure. This shows the structure of a gene in an animal or plant; bacterial genes do not have introns

A particular gene is found on just one of the strands of the DNA double helix. But other genes may be present on the other strand. Because of the large gaps between the functional parts of a gene, it is possible for genes to overlap, either on the same DNA strand, or more usually on opposite strands. Overlap of actual coding regions is unusual as there is obviously a severe constraint on the amino acids that could be coded simultaneously by the complementary strands running in opposite directions, but overlap of terminal sequences or introns, where there are few constraints on the possible sequence, is not uncommon.

Some genes encode RNA but not protein. These include the ribosomal and transfer RNAs already mentioned, but there is also a whole host of other RNAs such as microRNAs and long noncoding RNAs, which may be very numerous and whose functions are still uncertain.

Notwithstanding all these complexities, the gene of the molecular biologist is still a piece of DNA that has a function: whether protein coding, RNA coding or regulatory. As we shall see, the word 'gene' is now used in many life science specialities also to indicate non-coding and probably functionless sequences of DNA.

Each cell in an organism contains a nucleus with chromosomes which can be visualized down the microscope as short threads. Chromosomes are only visible during cell division. When cells are not dividing the chromosomes are not visible, but they are still there. Each is a single, very long, double helical molecule of DNA, containing many genes. Each chromosome comes in two copies, one from the mother and one from the father (Figure 7). Before a cell divides, the chromosomes are copied by replication of the DNA such that each double helical molecule becomes two double helical molecules. So each gene is present in each cell as two copies (one maternal and one paternal) before DNA replication and four copies (two identical maternal and two identical paternal) afterwards. When the cell divides, in the normal manner

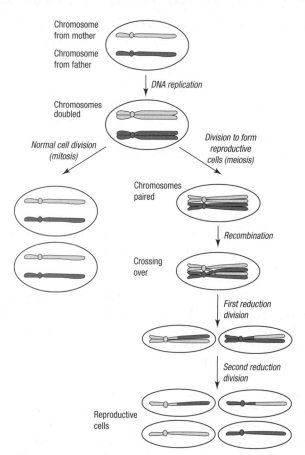

Chromosome from mother

Chromosome from father

DNA replication

Chromosomes doubled

Normal cell division (mitosis)

Division to form reproductive cells (meiosis)

Chromosomes paired

Recombination

Crossing over

First reduction division

Second reduction division

Reproductive cells

Genes

7. **Behaviour of chromosomes on cell division. On the left is shown normal cell division (mitosis). On the right is shown the formation of reproductive cells (meiosis). In males all four cells would become sperm, but in females only one of the four would become an egg**

which is called *mitosis*, the chromosomes themselves divide to separate the two identical DNA molecules, and they become partitioned equally to the nuclei of the two daughter cells. When reproductive cells are formed, a somewhat different cell division process called *meiosis* separates the maternal and paternal chromosome pairs, thus reducing the chromosome count in sperm or eggs to half of normal (Figure 7). Because maternal and paternal chromosomes segregate at random, each individual sperm or egg will contain a random selection from the two parents. The maternal and paternal chromosomes are also subject to recombination, whereby DNA molecules break and rejoin during the stage of chromosome pairing. So the degree of gene segregation is even greater than that achieved by the random assortment of whole chromosomes. When sperm and egg combine at fertilization, each supplies a single chromosome set and so the normal chromosome number is restored.

The complete sequence of all the DNA in a cell nucleus is called the 'genome'. With a few exceptions, the DNA in all cells of the body is identical so a sample of DNA from any tissue can be used to analyse the genome. Different individuals within a species have almost identical genes, although there are always some genes that differ slightly between individuals, consideration of which will represent much of the later part of this book. It is therefore possible to speak of 'the human genome' in which individual differences are ignored, and also an individual genome, in which every variant is specified.

Almost all animal and plant cells contain many small structures called mitochondria, responsible for the energy generation of the cell. These also contain DNA, which contains some of the genes encoding proteins from which the mitochondria are constructed. In the remote evolutionary past mitochondria were once free living microorganisms. This can be deduced from the fact that the sequences of the genes of their DNA are much more similar to those of modern bacteria than to their counterparts in the nuclear genome. At some stage the ancestor of mitochondria must have become incorporated into another cell to form a compound cell, or eukaryote, that was the common ancestor of all modern animals and plants (a process called 'endosymbiosis', Figure 8). Eukaryotes

Genes

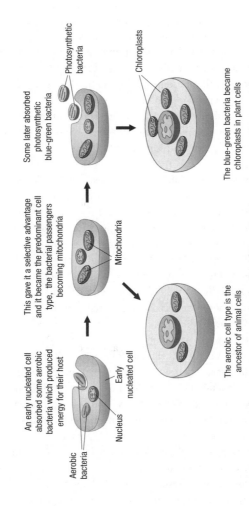

An early nucleated cell absorbed some aerobic bacteria which produced energy for their host

Nucleus   Early nucleated cell

Aerobic bacteria

This gave it a selective advantage and it became the predominant cell type, the bacterial passengers becoming mitochondria

Mitochondria

The aerobic cell type is the ancestor of animal cells

Some later absorbed photosynthetic blue-green bacteria

Photosynthetic bacteria

Chloroplasts

The blue-green bacteria became chloroplasts in plant cells

8. **Endosymbiosis. It is believed that mitochondria and chloroplasts originated as free living bacteria that were incorporated into cells in the remote evolutionary past**

26

have retained mitochondria with their own DNA ever since. As we shall see in Chapter 4, mitochondrial DNA turns out to be very useful for the tracing of ancestry through the female line, as any gene variants that become established are passed from the mother to all her offspring. The chloroplasts found in plants, which are responsible for photosynthesis, also contain their own DNA, and also resemble bacteria. It is thought that a second incorporation event occurred when the common ancestor to all plants engulfed the free living ancestor of all chloroplasts.

Genomics

In recent years the technology of genome sequencing has advanced enormously. At the time of writing it is possible to determine the complete sequence of the DNA from an individual human being very rapidly and for only a few thousand dollars. The '$1,000 human genome' is expected to be a reality very soon. Genomes of numerous animals, plants, and microorganisms have also been sequenced. This has generated, and will continue to generate, a mind-blowing amount of data. The human genome contains about three billion base pairs. Two different human individuals have genomes which are very slightly different so three billion more base pairs of information is generated for each individual who is sequenced. The storage, retrieval, and analysis of all this information has depended on massive computing power and has generated a whole new science of 'genomics'.

A typical genome contains much more non-coding DNA sequence than it does protein-coding DNA. There are methods for identifying genes, usually based on looking for regions that code for a long sequence of amino acids without any stop codons. These methods are not infallible and do not detect genes for which RNA is the final product, so the exact number of genes present in any complex organisms is still not certain. Animals have gene numbers in the range 13,000–30,000, so humans with about 20,000 are very typical in this regard. Plants tend to have rather more, in the range 25,000–45,000. Fungi have fewer at 5,000–11,000, and bacteria are the simplest organisms with 1,000–5,000 genes.

As indicated above, the ease of whole genome analysis has somewhat widened the concept of the gene itself. Only about 1.5 per cent of human DNA actually codes for protein and many genome-based studies do not look just at protein-coding genes but rather at the whole genome, including all of its non-protein coding expanses. Differences found between individuals may be correlated to disease susceptibility, or human migration, or forensic identification, but the DNA sequence variants examined are not necessarily functional genes. Instead they are 'markers' which are usually not genes at all in the molecular biology sense, although some may affect the activity of actual genes. As we shall see later, the genes of heritability studies and of evolutionary theory are even further away from hard molecular biology as most are not actually identified, and may even not exist at all. Their existence is inferred indirectly or simply postulated. So there are really many answers to the question 'what is a gene?' and to know the right answer it is important to understand the context of the question.

How genes make animals

In the 1980s and 1990s the mechanisms by which genes control embryonic development were discovered. These mechanisms are immensely complex, but the principles are quite simple (Figure 9). In essence different body parts or cell types are different because they express different sets of genes. The anatomy of an animal body is built up in a stepwise manner, starting with a single cell, the fertilized egg. In the first phase of development, called cleavage, the fertilized egg divides to form a sheet, or ball, of cells that are all very similar to one another. However there is always some asymmetry present. This often arises from the presence of substances originally localized in one region of the egg, which become inherited just by some of the daughter cells. The result of

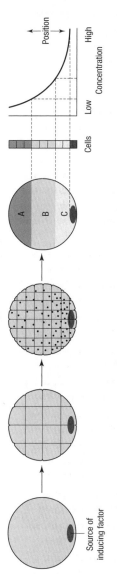

Different genes turned on at different concentration thresholds to create three cell populations with different properties.

Fertilised egg divides into cells. An inducing factor gradient is set up.

Source of inducing factor

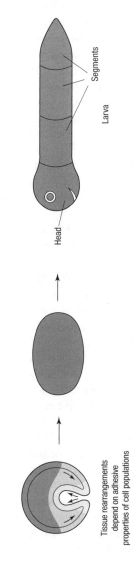

Tissue rearrangements depend on adhesive properties of cell populations

Head

Larva

Segments

9. **Development.** Indication of how genes act to form the anatomy of an animal during embryonic development. This is enormously oversimplified but indicates how a gradient of an inducing factor can cause activation of different genes in different regions. These genes then cause the regions to behave differently in terms of cell movements and growth. In reality there are many cycles of signalling, responses, and cell movements, not just one

this asymmetry is that the cells at one end of the sheet of similar cells become specialized to secrete a chemical signal called an inducing factor. The inducing factor spreads out to form a gradient of concentration and the cells of the sheet respond to different threshold concentrations of the factor by activating the expression of specific genes. These generally encode transcription factors, which are proteins that control the expression of other genes. Because of the gradient of the inducing factor, different combinations of transcription factors are produced in different regions of the early cell sheet. This process generates a simple spatial pattern consisting of several sorts of cell, each in a distinct region, and this constitutes the primary body pattern. It is more complex than the initial situation in which there were only two cell types, the signalling and the responding cells.

The primary body pattern is still very simple compared to a whole animal, but each of the primary regions then becomes subdivided in the same way, with a graded signal from an adjacent region leading to formation of new boundaries of gene expression. With five or six cycles of subdivision a very complex anatomy can quickly be built up. The signals involved in these processes, and the genes turned on in response to them, have now been largely identified. The process described is called regional specification, and other processes which are essential to development occur simultaneously. Cell movements enable a single sheet of cells to become multilayered, and for specialized regions to move as a cell mass into neighbouring areas. Growth, driven by cell division, increases the overall size of the embryo, and may also generate differences of size between parts. The processes of regional specification, cell movement, and cell division are quite well understood and most of the key genes that control these events have been identified. Much less well understood is how the timing of development is carried out. We still do not understand why a mouse develops so much faster than a human, or even whether there is or is not any 'master clock' controlling development.

Although it is generally considered that the same genes are present in all cells of the body there are some exceptions to this. One important one is the cells of the immune system responsible for making antibodies and cell-based recognition of infection. The large gene repertoires responsible for these processes are based on an inherited set but are greatly augmented in variety by mutations occurring during development. This is why even identical twins can have different antibody responses. There are other mutational events, particularly common during development of the nervous system, which may help to cause some of the differences between individuals.

The molecular basis of development was worked out by studying embryos of a variety of experimental animals, including frogs, mice, fruit flies, and nematodes. One remarkable result from this work was the insight that the mechanism of development of different animals had a lot in common. Essentially the same genes and the same inducing signals are used to build a frog as a human, and even the fruit fly uses many of them. Of course they are not exactly the same, but when we talk of 'the same' gene in a human and a fly, we mean DNA sequences that share many bases, and generate proteins with similar biochemical functions and specificities. Such genes will often have the same name. The similarity arises because of evolution. One gene in the common ancestor of all animals gave rise to similar genes in humans and flies over 600 million years of evolution later. Although many mutations arise and the bases present at many positions are altered, not everything has changed and it is still possible to discern the relationship between even quite widely diverged genes.

The same general principles apply to plant development, although plant cells do not move around like animal cells and so the pattern that is laid down in a growing shoot or root apex remains the same thereafter. Also the actual genes and inducing signals involved in plant development are quite different from those found in animals, indicating that animals and plants evolved from different single-celled ancestors.

There has been some debate about whether it will ever be possible to 'compute' the final appearance of an organism from the DNA sequence of its genome. For a completely unknown organism, I suspect that the answer will always be 'no'. Although the organism is, in a sense, fully specified by its genome, the execution of development involves various higher-level processes, such as cell-signalling events and cell movements, which need to be studied and understood at their own level. It is true that if one were presented with the sequence of an unknown genome, it is always possible to compare it to known genomes and say what sort of organism it is because of its affinities. But what about organisms which may yet be discovered on other planets or other stars? They may have the same DNA as we do, and perhaps even the same genetic code and types of protein. But they will have a totally separate evolutionary history. It will almost certainly not be possible to say what kind of organism an alien genome encodes, just from knowing the DNA sequence, and especially if it contains more than a few thousand genes.

Despite the unlikelihood of being able to predict a whole organism from its genome, it is already possible to make accurate predictions of the developmental effects of removing or adding specific genes in well studied organisms. The technology for doing this exists in several experimental animals such as frogs, fish, or mice, and it is often possible to remove or add the activity of a gene from a specific region of the animal and to do so at a specific time in development. Such experiments make up a large part of contemporary developmental biology research. In terms of the ability to predict the effects of such changes, we can claim that we do now have a good understanding of the genome and how it controls development.

The human genome

The complete sequencing of a typical human genome was started in the late 1990s and achieved in 2003. At the time the technology was not nearly as good as it is now so the exercise, conducted by the US National Institutes of Health with some international partners such as the UK Wellcome Trust, cost nearly $3 billion. Analysis of the total sequence was a very complex exercise and is still going on today in terms of refinement of bioinformatic analysis. It showed that the genome of human beings contains about three billion base pairs of DNA. The process of gene identification is complex and it is hard to guarantee that all have been found, but the present count is about 20,500 protein coding genes. Perhaps surprisingly this gene number is no larger for humans than for most other types of animal, including some apparently quite simple animals, and it is smaller than found in many plants. A figure of 20,500 proteins has been considered by some to be 'not enough' to make up all the exquisite complexity of a human body. However, many of these proteins exist in multiple forms because of different intron removal patterns and chemical modifications after the initial synthesis. Moreover those genes involved in development are subject to very complex regulation from numerous regulatory elements in the DNA. In addition, the developmental identity of body parts and cell types is specified by combinations of transcription factors (proteins that control the activity of other genes) and the available combinations of just a few hundred such proteins make up a very large number, for example comfortably exceeding the number of nerve cells in a human brain, so the genome certainly does not lack resources for the job it has to undertake. The number of RNA-only coding genes is not so well established, but some estimates indicate that they may be very numerous, adding significantly to the overall complexity of the human genome.

The genome shows every sign of having arisen through a process of evolution. It seems unlikely that an 'intelligent designer' would have used so much DNA for the non-functional repeated sequences which make up about 45 per cent of the total genome. The existence of repeated sequences can however be explained through insertion, movement, and copying of transposable elements in the course of evolution. Many of these elements used to be viruses in the remote past. Once they enter and integrate into a genome the nature of DNA guarantees that they will be replicated forever, although the functional parts responsible for mobility are usually lost after a small number of generations. It is also unlikely that a 'designer' would have chosen to put just a few genes into the self-replicating mitochondria where they suffer from oxidative damage and poor DNA repair, thus generating many mutations, some of which are lethal to the whole organism. The human genome also includes about 15,000 'pseudogenes' which are not expressed but which have a clear sequence similarity to functional genes. These arose from gene duplication followed by accumulation of mutations in one copy.

All of this kind of information, obtained from genomic sequencing, indicates that the human genome shows every sign of having had a long past history of evolution, taking place partly by means of natural selection and partly by means of genetic drift, concepts discussed further in Chapter 6.

Chapter 3
Mutations and gene variants

So far we have been considering the nature of 'the gene' in the sense used by molecular biologists: the agent that is responsible for transmission of information from one generation to the next and which controls the development of the embryo. Genes in this sense determine that a mouse embryo shall always develop as a mouse and not as a dog or jellyfish or buttercup. Between any two humans, about 99.9 per cent of their genomes are identical in deoxyribonucleic acid (DNA) sequence. But about 0.1 per cent differs. When we hear of 'genes for cancer', 'Northern European genes', or 'genes for high IQ', it is not really genes but the *differences* between genes that are being referred to. Biologists refer to different versions of the same gene as 'alleles', but in this book I shall use the term 'gene variant'. Such differences may occur anywhere in the genome although between any two individuals they make up only a very small fraction of it. The place where a gene occurs in the genome, which may be occupied by the normal version or by a variant, is known as a 'locus'.

Ultimately all gene variants originate as mutations. Most variants in the genome of any given individual are not new mutations but have been inherited from previous generations, however, they did originate at some time in the past as new mutations. Because of this the terms 'gene variant' and 'mutation' are often used interchangeably. We shall consider later the nature of genetic variation within populations and focus for now on mutations themselves.

Mutations

Since the original discovery of mutations by de Vries, it has been known that alterations of the genome of an individual can occur suddenly. Any of numerous possible types of chemical change to the DNA double helix can be a mutation. Mutations can occur in any cell of the body but in order to be inherited they must occur in the DNA of the reproductive cells that are destined to form sperm or eggs. Mutations occurring in other body cells may be important, for example they may contribute to the development of cancer, but they cannot be inherited by the next generation. Because most changes in the DNA sequence are introduced during DNA synthesis, mutations are much more common in cells that are dividing compared to those that are quiescent. Although the occurrence of mutations is random in relation to the consequent biological effects, it is not random throughout the genome. There are some 'hot spots' which may reflect ease of access of the DNA to reactive chemicals or to the enzymes that repair DNA damage.

The simplest kind of mutation is the change of one base to another. This can occur through purely random chemical processes. For example the base guanine normally occurs in the 'keto' form. But it can occasionally flip to the less stable 'enol' form (Figure 10). The enol form of G pairs with T rather than with C. So when DNA synthesis occurs, a T is inserted opposite the enol-G. At the next round of replication this T will be paired with A, so a permanent and heritable change will have occurred. Single base changes can also arise from chemical damage which may be caused by reactive chemicals called mutagens. For example ethyl nitrosourea is commonly used in biological experimentation. This adds an ethyl group (C_2H_5-) to position six of G which makes it pair with T instead of C. There are many types of mutagen present

10. Keto-enol. Spontaneous changes of chemical structure which occur in DNA bases and can cause incorrect pairing of bases during DNA replication

at low levels in our environment, for example generated by smoking of cigarettes, chemical changes in food during cooking, or produced by moulds growing on food. Depending on the precise chemistry involved, in some cases reaction of a mutagen with a base causes mispairing, while in others it causes a repair process to start which involves removal of part of the DNA strand and its resynthesis to close the gap. This process may introduce errors, which are heritable mutations. Ultraviolet (UV) light, encountered in sunlight or at tanning salons, can join together adjacent thymines, which are then excised and repaired, again with the possibility of error. This is why UV light is a cause of skin cancer although not of mutations heritable by offspring since its effects are largely confined to the skin. Ionizing radiation (medical X-rays or background radioactivity) produces double stranded breaks, which are harder to repair and lead to more errors when they are repaired. Ionizing radiation is highly penetrative into the body so can cause heritable mutations in reproductive cells as well as in the rest of the body. The presence of multiple reiterated sequences in the DNA also makes mutation likely. During DNA replication, a change of register can occur such that the number of repeats becomes increased or decreased. Although such repeats are not usually present in protein coding genes, changes can have effects on the activity of neighbouring genes. The biggest changes to DNA can be so large as to affect the visible structure of chromosomes, and can be seen down the microscope, for example as a transposition of a chromosome segment to another chromosome. But the vast majority of mutations are of molecular dimension and cannot be seen down the microscope.

Any of the above types of mutation may or may not alter a protein. To do so it needs to be in a protein coding sequence and it needs to alter one or more base triplet from one amino acid to another. In developmental biology, mutations are classified as loss-of-function or gain-of-function. A loss-of-function mutation may prevent the production of the protein, or reduce its metabolic or functional activity. So long as the other copy of the gene is still in working order, the overall activity of the protein cannot be reduced below 50 per cent. This is often enough to maintain normal function and so loss-of-function mutations usually only show effects if both copies of the gene are affected. In general, gene variants that only show an effect when both copies are affected are known as *recessive* to the normal version. In the converse case, where presence of the variant in only one copy has a visible effect, the variant is called *dominant* over the normal version. If 50 per cent of gene activity is not enough for normal function, one copy of a loss-of-function variant will cause an effect, and it will be classified as dominant, although the lack of both copies will have a much stronger effect. Gain-of-function mutations come in several types, all of which are usually genetically dominant. Many represent 'constitutive activity' of a protein. For example a cell surface molecule that is normally stimulated by a hormone may be mutated so it is active all the time, whether the hormone is present or not. Some are 'dominant negative' and produce a toxic product that inactivates the product of the normal gene copy, for example by forming inactive molecular complexes with it. Both these types of mutation are genetically dominant because the effect will be exerted whatever the nature of the other copy of the gene.

Recently the use of whole genome sequencing has made it possible to calculate the mutation rate of human beings fairly accurately. It is about 1.2×10^{-8} per nucleotide per generation. In other words any individual is likely to carry about 63 new mutations. These mutations arise during development of the sperm or eggs of the parents, and here there is an interesting asymmetry. Female reproductive cells only divide during embryonic life and a girl is born with her entire lifetime supply of eggs already formed. But in males sperm are produced throughout life from stem cells, so by the time it is formed the average sperm has passed through many more DNA replication cycles than the average egg. Because mutations are established mostly during DNA replication this means that more mutations are inherited from fathers than from mothers. A recent study in Iceland indicated that each individual inherited about 14 new mutations from the mother and 55 from the father. Moreover individuals have about twice the number of new mutations if the father is aged 40 compared to 20. There is also a correlation with maternal age since people tend to marry partners of a similar age, but the effect is thought to arise mostly from the accumulation of mutations in the sperm-forming stem cells of the father.

Gene variants in populations

Since all cells of each organism contains a chromosome set from the mother and one from the father, they all contain two copies of each gene. If these are alike the individual is said to be *homozygous* and if they are different the individual is *heterozygous* for that particular locus in the genome. If a gene variant is recessive to the normal form, in other words its effect is not manifested in the presence of the normal form, the heterozygotes are known as 'carriers' because they 'carry' the trait to future generations without displaying it themselves. When two carriers of a recessive variant mate, 25 per cent of offspring will be homozygous for the variant and will display the effects, 50 per cent will inherit one copy of the variant and be carriers themselves, and the other 25 per cent will have two normal copies of the gene, just as shown in Figure 2 relating to the colour of peas.

Because gene variants are mostly inherited from previous generations, the frequency of a particular gene variant in the population is generally much higher than the new mutation rate leading to formation of the variant. The only exception is if the mutation is completely lethal and dominant, in which case all individuals acquiring it will fail to reproduce, and the prevalence in the population will equal the mutation rate (about 0.0001 per cent per gene). A gene variant that is completely lethal but fully recessive will be present at about 0.1 per cent individuals because this is the frequency at which new mutations are balanced by the non-reproduction of individuals who inherit two copies from their carrier parents. Gene variants that are not completely lethal can be expected to be present at higher frequencies than this. In fact many are completely neutral, with no effect on the reproductive performance of their carriers. In this case their frequency is governed just by chance sampling from one generation to the next and can range from zero to almost 100 per cent. If the frequency of a gene variant actually reaches 100 per cent it becomes, by definition, the normal form (this process is shown later in Chapter 6).

In addition to these considerations based on the new mutation rate, there are several other factors that may affect the frequency of a particular gene variant in the population. This issue is particularly significant for disease-causing mutations that seem to be present at a higher level than would be expected from the balance between mutation and selection. For example, sickle cell anaemia is a common disease that affects the structure of haemoglobin, the protein of red blood cells responsible for carriage of oxygen. Sickle cell disease is caused by a single nucleotide change which alters the amino acid glutamic acid to valine at position six of the haemoglobin beta protein. In heterozygotes, 50 per cent of the protein is still normal and there are few deleterious effects. But in homozygotes all of the haemoglobin beta is in the abnormal form and tends to form molecular complexes, reducing function and altering the shape of the red blood cells to the characteristic 'sickle' form. The disease has multiple pathologies. Every so often the abnormal cells occlude blood vessels leading to painful crises in which organ damage can occur. Patients often lose the function of their spleen early on and are vulnerable to infections. Even today, for those fortunate enough to receive optimal treatment, the life span is considerably reduced. In the past it would have been very short and few sufferers from the disease would have reproduced. For a disease that has minimal effects on heterozygotes but largely prevents reproduction in homozygotes, we might expect a gene frequency of about 0.1 per cent and a disease prevalence of about 0.0001 per cent. However it can be much higher than this. In parts of west Africa the gene frequency is about 20 per cent and the prevalence of the disease about 4 per cent. This is because the heterozygous condition itself confers a selective advantage, which is a degree of resistance to malaria. The malarial parasite lives in red blood cells but in the carriers these cells are somewhat more fragile than usual and often rupture before the parasite can reproduce. This situation, where the heterozygote is fitter than either homozygote, is one commonly encountered cause for high frequencies of otherwise deleterious gene variants.

Another possible cause for high variant frequency is historical. For example, Ashkenazi Jews have a higher than expected frequency of a gene variant causing Tay-Sachs disease, which involves an inability to degrade certain types of lipid that are components of the cell surface, and leads to an early death due to the accumulation of these lipids in neurons. Among American Ashkenazis the proportion of carriers is about 3.3 per cent and that of affected births about 0.027 per cent. In the case of Tay-Sachs disease there is no known advantage for heterozygotes and it has been speculated that the reason for the relatively high prevalence is the severe historic contractions of the Jewish population due to suppression of the Jewish rebellion by the Romans in the year AD 75, and again during pogroms in the period AD 1000–1400. If a small number of individuals give rise to a large population, any gene variants that they happen to carry will be overrepresented in the later population. Eventually an evolutionary equilibrium will be re-established by natural selection against the variant, but this will take a very long time for a deleterious recessive gene variant because the vast majority of the mutant genes are located in symptomless carriers.

Cystic fibrosis

An example of a relatively common, recessive, loss-of-function, genetic disease is cystic fibrosis. This arises from mutations causing loss of activity of a protein, unimaginatively named cystic fibrosis transmembrane conductance regulator (CFTR). The gene is a particularly large one and is therefore quite susceptible to mutation. The protein is responsible for transport of chloride ions across cell membranes. CFTR is normally present in the cells of several tissues including particularly the lungs and the pancreas. The disease therefore manifests itself in various ways. In the lungs, the decreased transport of chloride ions thickens the secreted mucus and reduces its rate of clearance. This leads to difficulties in breathing and also increases the risk of respiratory infection. In the pancreas the effectiveness of secretion of digestive enzymes is greatly reduced, leading to damage to the organ and problems with digestion. In addition, male sufferers are usually infertile due to the failure of the tubes conducting sperm to the exterior to develop in embryonic life.

There are many different gene variants giving rise to cystic fibrosis, but all of them act by reducing the function of the CFTR protein. The most common variant worldwide is the loss of three nucleotides that cause the absence of one amino acid, phenylalanine, from position 508 in the protein (the mutation is called ΔF508). Different mutations cause different degrees of loss of function and so different degrees of severity of the disease. For example patients with two copies of mild mutations which preserve 5 per cent of protein function still display male infertility but do not have major effects on the lungs or digestive tract. Those with two copies of mutations preserving more than 10 per cent of normal activity do not show any significant abnormality at all.

The frequency of the cystic fibrosis gene variants in the population varies considerably in different parts of the world. The disease is rare in Asia but relatively common in Europe. For example, in the UK, about 1/25 people carry one copy of one of the mutations. This means about one sperm or egg in 50 contains a cystic fibrosis causing gene variant, as the reproductive cells inherit the paternal or maternal chromosome at random when they are formed. With random mating this means that about 1/(50×50) or 1/2,500 individuals are affected by the disease. In the UK, 75 per cent of the cystic fibrosis variants are the ΔF508 type referred to above. The frequency is remarkably high for such a serious disease variant given that until very recently no individual with the severe (ΔF508) type of cystic fibrosis would have survived to reproductive age. It has been speculated that the heterozygous state gives protection against some infectious diseases common in Europe that caused significant mortality. But no firm conclusion is yet possible on this.

The cystic fibrosis gene was identified and cloned as early as 1989. At the time it was hoped that it would be possible to introduce the normal gene into the airways of patients to cure the disease. Unfortunately after 20 years of effort this has still not proved possible due both to inefficient gene transfer into airway cells and to immune responses against the viruses used for gene delivery. Nor has the identification of the gene responsible enabled development of a new drug therapy targeted at the gene product. The more conventional treatments on offer have, however, improved a lot in recent decades. In the 1950s patients would die as babies, but now, with good physiotherapy, control of infection, and nutritional support, median survival extends to mid-adulthood. However, the disease continues to cause considerable suffering and the treatment is very arduous. When an effective gene therapy procedure is finally developed it will be most welcome.

Haemophilia

Haemophilia is another well-known genetic disease. Its inheritance differs from that of cystic fibrosis in that only males are usually affected. There are two main types of the disease: haemophilia A is due to the lack of a protein called factor VIII, and haemophilia B to the lack of a protein called factor IX. Both of these proteins are normally found in the blood and are needed for the normal blood clotting mechanism. So individuals suffering from haemophilia have little or no ability to clot their blood. This causes problems following any injury and can often lead to internal bleeding. Bleeding into joints causes arthritis and bleeding into the brain can lead to strokes and sudden death.

It is often considered that if a condition is genetic this means that it is incurable. But this is not so. Haemophilia can be successfully treated by periodically infusing the missing protein, factor VIII or IX, as appropriate. However, this is expensive; at the time of writing the annual cost is about $300,000. Before the era of modern treatment most haemophiliacs would die in childhood, but now the life expectancy is approaching the normal. Factors VIII and XI used to be extracted from pooled human blood and this unfortunately meant that many haemophiliacs became infected with AIDS when the epidemic started in the 1980s and many died as a result. Since the introduction of factors VIII and IX made by recombinant DNA methods this has not been a problem. Experiments have been conducted on gene therapy in which a normal copy of the gene is introduced into the liver, which is the organ that normally produces the two proteins and it is likely that this will eventually become the preferred treatment.

Many different mutations in the genes for factors VIII and IX can cause haemophilia. Because the effects on the protein vary from partial to complete loss of function there is considerable variation in the severity of the disease depending on which mutation is responsible. About 1/3 cases are due to new mutations and about 2/3 to inheritance of the gene variants from the parents. The incidence is about 1/10,000 births for haemophilia A and 1/50,000 for haemophilia B. Why does haemophilia occur almost entirely in males? This is because both genes lie on the X chromosome. In humans, as in all mammals, females have two X chromosomes and males have one X and one Y. If a female has one copy of a loss-of-function haemophiliac gene variant she will make 50 per cent of the appropriate clotting factor, which will be sufficient for most purposes, and so she will be an unaffected carrier. Her daughters have a 50 per cent chance of inheriting the mutant X chromosome and if they do they will likewise be carriers. Her sons will also have a 50 per cent chance of inheriting the mutant chromosome but those that do inherit it will all have the disease. This is because the other sex chromosome in males is the Y chromosome from the father and this does not contain either of the factor VIII or IX genes. So the male with a haemophilic variant on his X chromosome will be defective in clotting factor. It is not impossible for a female to have haemophilia but to do so she will need to inherit two copies of the mutation. This means she needs to be the daughter of a haemophiliac man and a carrier woman, which is obviously a fairly rare situation. Of course a haemophiliac father cannot pass the disease to his sons because they will necessarily acquire his Y chromosome and not his X chromosome. Because most of the genes found on the X chromosome are not present on the Y, there are many other sex-linked disorders due to mutations in these other genes. All are much more common in males than females.

The most famous carrier of haemophilia was Queen Victoria. She carried a new mutation for haemophilia B which presumably arose in her mother's egg-producing cells. She passed the disease to her own son Leopold who passed it on to his daughter Alice. Alice's son Prince Rupert bled to death after a car accident. Leopold himself died at the age of 30 after a fall. Victoria also passed the haemophiliac gene variant to her own daughter, also named Alice, who was the grandmother of Tsarevich Alexei Nikolaevich of Russia (Figure 11). He had the disease although he was executed before he died of it. In 2009, sequencing of DNA recovered from the skeletons of the Tsar's family confirmed that Alexei suffered from haemophilia B and that his mother, Alexandra Feodorovna, and sister, Anastasia, were both carriers. Victoria's daughter Beatrice was also a carrier, with the result that two princes of Spain, Alfonso Prince of Asturias and Infante

11. **Romanov family. The seated man is Tsar Nicholas II, the last tsar of Russia. The crowned lady behind him is Empress Alexandra Feodorovna, carrier of haemophilia B, which was inherited from her maternal grandmother Queen Victoria. The boy is Alexei Nicolaevitch, their son, who suffered from haemophilia**

Gonzalo, who were her grandsons, were affected and both died by bleeding to death after car accidents.

Achondroplasia

A well-known example of a dominant, gain-of-function condition is achondroplasia, which is the most common form of human dwarfism. The short stature is mostly due to abnormally short arms and legs, and there are some other skeletal abnormalities, for example usually a prominent forehead and bowed legs. Achondroplasia is a developmental abnormality arising from a mutation in the gene for a particular growth factor receptor known as fibroblast growth factor receptor 3 (FGFR3). The substances called fibroblast growth factors (FGFs) are very important in embryonic development and they act to stimulate FGF receptors which are molecules on the cell surface that send a biochemical signal into the cell to regulate expression of many genes. The gene variants causing achondroplasia cause the receptor to be active all the time rather than only when FGF binds to it. The FGFs have many functions but in skeletal development they normally act to reduce growth of cartilage and bone. So if the receptor is overactive, the inhibition is greater than normal and this leads to the abnormalities, especially the shortening of the long bones of the limbs. Because the defect is caused by an overactive protein, the abnormal variant is genetically dominant over the normal form and only a single copy of the gene variant is needed to cause the condition.

The incidence of achondroplasia is about 1/20,000 births. Unlike cystic fibrosis, where most cases arise from the coming together of gene variants already present in the population, about 80 per cent of cases of achondroplasia arise from new mutations. In nearly all of these a point mutation at nucleotide 1138 of the *FGFR3* gene causes a glycine to arginine substitution at protein position 380. Interestingly it has been found that virtually all of these new mutations arise in the father. This is partly because, as indicated above, all mutations are more likely to come from the father because of the large number of divisions of the sperm-forming stem cells, and the consequent continual risk of mutation at each DNA replication. There is indeed a strong correlation between the risk of achondroplasia and paternal age. However, this bias alone does not account for the lack of mutations in the mothers. In the case of constitutive mutations of FGF receptors it is possible that these confer some advantage to the sperm-forming stem cells, in terms of division rate, or even to the sperm themselves, in terms of increased motility and survival. FGF receptors have a wide variety of functions and these normally include stimulation of cell division, although, as we have seen, in cartilage FGFR activity reduces cell division.

Some cases of achondroplasia are inherited from achondroplastic parents. The condition does not prevent reproduction so affected individuals do have children. For a dominant gene variant the probability of passing to the offspring is 50 per cent, as the reproductive cells may carry either the normal or the mutant version of the gene at random. Achondroplastics do often marry each other. If both parents have one copy of the mutant variant then the risk of an achondroplastic child is 75 per cent. However, of these 1/3 (25 per cent of total offspring) will have two copies of the mutant gene. Compared with those having one copy, those with two copies are much more seriously affected and always die in utero or shortly after birth.

Note that the conventional idea of a genetic disease is something inherited from the parents and grandparents. But since most cases of achondroplasia arise from new mutations, it is a disease that it is not usually inherited in this sense although it is certainly genetic.

Holt-Oram syndrome

Unlike the other diseases mentioned here, this one is unlikely to be familiar to readers. It was first described in 1960 and is rare, with an incidence of about 1/100,000 births. But it exemplifies a number of interesting features. It is a dominant developmental disorder which affects the arms (but not the legs) and the heart. The bones of the wrist are usually abnormal, the long bones of the arm may be missing and extra digits may be present. In the heart there are often holes in the wall separating the right and left sides. The mutations occur in the gene encoding a transcription factor called TBX5. Transcription factors are proteins that control the activity of other genes in a specific manner and about 200 of them are of critical importance in embryonic development. The genes for the transcription factors themselves become active in response to the graded inductive signals that control the subdivision of the early embryo. They then activate many other genes. One transcription factor controls many genes, and each gene is usually regulated by several transcription factors so there is a lot of complexity to the 'wiring diagram' of the developmental programme. Studies by developmental biologists identified *TBX5* as being necessary for the normal development of the forelimbs (but not the hindlimbs), and the heart (Figure 12). Once you know these facts about development it is no longer a surprise that individuals carrying mutations in the gene have defects both in the forelimbs and heart, but before this was known the clinical syndrome was completely baffling. Several different mutations in the *TBX5* gene can all cause Holt-Oram syndrome and most of them reduce or abolish the function of the protein. This is a case where the genetic dominance arises because 50 per cent of normal gene activity is not enough, so inheriting just one copy of the inactive gene from a parent leads to the condition. Holt-Oram syndrome is not lethal and a proportion of cases arise by direct inheritance of the gene variant from an affected parent. Quite a few also arise from new mutations. It is believed on the basis of studies of this same mutation in the mouse that an individual inheriting two copies of the mutant gene and thus lacking all *TBX5* function would most likely die early in pregnancy.

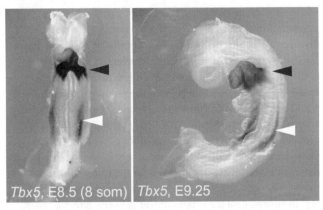

12. *Tbx5*. Expression of the gene *Tbx5* in the early mouse embryo. The dark patches represent regions of gene activity visualized by a technique called in situ hybridization. Note the activity both in the developing heart (black arrowheads) and limbs (white arrowheads). E8.5 and E9.25 indicate the gestational age in days

Genetic predisposition and cancer

The examples of genetic disease considered so far are due to a single mutation in one gene. In such cases the inheritance of the abnormal gene variant follows simple Mendelian rules and the level of risk can be predicted accurately by genetic counsellors if they know the genetic makeup of the parents. However, most human diseases that have a hereditary component are not nearly so simple. Usually gene variants at several positions in the genome each contribute a small or modest increase to the overall risk of a deleterious outcome. We shall consider in Chapter 5 just how it is known that gene variants are involved in such cases. For now we shall consider an example of great interest and importance: predisposition to cancer.

Cancer is to a large extent a genetic disease but it is not usually inherited from parents. It is genetic in the sense that significant damage to the DNA of a cell is necessary for that cell to give rise to a cancer. But most cells are those making up the various functional tissues of the body, and are not the reproductive cells, which form the sperm or eggs. Only mutations in reproductive cells can be passed to offspring. Mutations occurring in normal body cells cannot be passed to offspring but they are passed to daughter cells on cell division and can still have consequences for the individual.

Our bodies contain a number of dividing cell populations, in particular the stem cells forming the blood, the skin, and the lining of the intestine. Every time the DNA is replicated, prior to cell division, there is a risk of mutation. Certain types of mutation cause cells to become cancerous, that is, to divide without restraint and to spread to other parts of the body. Eventually the growth of primary or secondary tumours causes sufficient mechanical damage to organs to cause death. The nature of the mutations that cause cancer has been intensively studied and falls into a number of classes. These include mutations that stimulate cell division, that reduce sensitivity to inhibitors of division, that suppress cell death, that encourage the growth of blood vessels in the vicinity, and that disable the mechanisms of DNA repair, making the establishment of new mutations more likely. Most types of cancer increase in incidence very steeply with age, proportional to about the 4th or 5th power of age. This is because it takes several specific mutations acting together to make a cell cancerous and the probability of collecting five to six such mutations in one cell from randomly occurring mutations will naturally rise with time elapsed. Most cancers arise because of this accumulation of mutations, and most of these arise from purely random causes: spontaneous chemical changes, natural background radioactivity, oxidative damage consequent on respiration, and so on. This is why 1/3 of us will die of cancer and if new miracle treatments came along to abolish heart disease, stroke, and pneumonia, our gain of lifespan would be only modest because the incidence of cancer increases so steeply with age that we should all die of cancer within a few more years.

There are some relatively common gene variants that are important in predisposing to cancer but do not cause it with the same predictability with which, for instance, the lack of *CFTR* leads to cystic fibrosis. Examples of such predisposing gene variants are mutations in the genes *BRCA1* and *-2* (pronounced 'braka') which increase the risk of breast and ovarian cancer. The overall lifetime risk of breast cancer for a woman is about 12 per cent and this is increased to 50 per cent or more when she carries a mutation of *BRCA1* or *-2*. For ovarian cancer the normal lifetime risk is about 1.4 per cent and this is increased to maybe 40 per cent if the woman carries a *BRCA* mutation. There are also significant increases in risk of prostate cancer, for men, and of colon cancer for both sexes.

If there were no DNA repair the mutation rate would be very much higher than it is. BRCA1 and -2 are both proteins that are involved in the repair of DNA damage and the variants impair the effectiveness of the process, thus increasing the mutation rate. There are numerous variants of *BRCA1* and *-2* most of which give rise to a reduced function protein. This might suggest that they should be recessive, since there will still be the good copy of the gene present on the other chromosome. However at the level of the whole person they are actually dominant in the sense that possession of one bad copy confers an extra risk of cancer.

One way of looking at the process is to consider that, say, six specific mutations all need to occur in one cell for it to become cancerous. All the cells of the body are experiencing mutations all the time but, since the body contains more than 10^{13} cells altogether, to collect six specific ones in the same cell takes a long time, and may never happen. However if the woman already carries a loss-of-function variant in all cells of her body because of inheritance, then instead of six mutations you now need only five. The odds against this happening for any individual cell are still huge, but for the whole body it means a significant increase in risk. This is why the *BRCA* mutations are genetically 'dominant' at the level of the whole body, at the same time as they are recessive at the level of the single cell. In fact many of the established breast cancers in women carrying *BRCA* mutations have lost the good copy of the gene as well, leaving no BRCA function at all in the cells of the tumour. Once this has occurred the mutation rate ramps up considerably within the tumour and increases the risks of further malignant transformation leading to more rapid growth or to spread to other parts of the body.

Some cancers are known to be associated with infection by particular viruses. In these cases the virus often introduces a gene which mimics one of the mutations that is required for development of cancer. But not every virus-infected cell becomes cancerous because virus-induced cancers also depend on the occurrence of additional mutations in the host.

The example of the *BRCA* genes is a very important one because of the fact that breast and ovarian cancer are both quite common and so an increase of risk is a serious matter for the individuals concerned. It might seem that we should screen everyone for possession of these mutations. Unfortunately there are many different mutations in both genes and so it is difficult to screen for all of them. Even if the genes were sequenced it is not possible to be sure there are no mutations outside the coding region that depress gene activity. It is however possible to screen an individual within a family with high cancer incidence and already known to possess a *BRCA* mutation, for inheritance of this particular mutation, and this is now done in many places.

More complex genetic variation

The examples given in this chapter are of genetic variations in the human population in specific genes, with known functions, where the mutations have clearly defined consequences. Understanding each of these examples has been a triumph of genetics, and they illustrate a number of principles about what genes are and how they work. However most genetic variation is not of this type. In Chapters 5 and 6 we shall look at several situations where there appears to be a hereditary component in important human or animal characteristics, but where few or no specific genes have been identified. It may be that the simple model of single genes controlling specific aspects of the organism, exemplified by the examples above, is the exception rather than the rule, and that heredity in general is of a more quantitative character, involving thousands of genes each exerting only a small effect.

Chapter 4
Genes as markers

Although some mutations are certainly deleterious, and a very few are advantageous, the vast majority of changes arising in deoxyribonucleic acid (DNA) do not significantly affect the reproductive fitness of the organism. The genetic variation dealt with in the previous chapter was a specially selected subset of examples that do affect specific genes and lead directly to disease. But most genetic variation does not affect gene function or activity. This is because most changes to the DNA sequence lie outside the genes, in the large majority of DNA that does not code for proteins. Even variation occurring within the regions of sequence coding for proteins may not have an effect. Because there are 61 nucleic acid triplets encoding 20 amino acids, several triplets are 'synonymous', encoding the same amino acid, and a mutation converting one triplet into a synonymous one will not affect the protein structure. There has been much debate about the extent to which genetic variation is genuinely without effect on the organism ('neutral') and to what extent it has small effects which could be significant over long time scales. Some authors have always considered that the number of variants carried by every one of us is so large as to represent a crushing reduction of reproductive fitness if they all had deleterious effects. Other authors are always coming up with reasons why variation, even if it is not in actual genes, may have small effects on fitness through indirect effects on gene function. This debate is still not resolved and in particular it remains unclear whether large areas of the genome do or do not encode non-translated RNAs and whether these RNAs, if they exist, have any functions. However, to a first approximation it is safe to conclude that a high proportion of genetic variation is neutral or nearly neutral.

In genetics the word 'locus' indicates a place in the DNA, regardless of which particular gene variant is present. Each variant exists because a mutation occurred at some time in the past in a single reproductive cell of a single individual. For a mutation that is neutral in its effects it is a matter of pure chance whether it or the normal version of the locus becomes inherited by the next generation. Chance will further dictate whether it becomes more common in subsequent generations, or whether it will become less common and eventually be lost from the population altogether. Of the large number of mutations that occur all the time in reproductive cells, a very few will eventually spread right through the population by chance and become the normal version of the gene.

Although most genetic variation probably does not affect the organism or its reproductive fitness, it is still of enormous interest. Particularly for the human population the ability to look at individual variation has generated several new sciences. Notably it enables the identification of individual people, useful in forensics, the establishment of paternity, and other information about family relationships. It also provides some evidence about the migration of human populations in historic and prehistoric times. Genetic variation has also enabled biologists to examine the thorny issue of human racial differences and establish the degree to which there is any biological basis for perceived race.

A word that is quite often used to refer to a gene variant in this context is 'marker'. Rather than refer to 'genes' that distinguish individuals or races, I shall refer to markers, as most of them are without known function. This chapter then deals with markers:

entities that are real in the sense that they can be identified by DNA sequencing machines, but are not necessarily 'genes' in the strong sense of the protein coding genes having specific functions like those described in the previous chapter.

Forensic identification

Any of the many types of genetic variation could in principle be used for individual identification. However, methods designed for use in criminal proceedings have to be very robust and reliable, and a high degree of standardization is necessary to create national databases and to make comparisons possible on a large scale. For these reasons forensic science has focused on a type of marker known as the simple tandem repeat (STR). STRs are very simple DNA sequences, three to five nucleotides in length, which are repeated several times as copies lying next to each other. At a particular locus there might typically be 5 to 30 copies arranged end to end (i.e. in tandem). The number of repeats at a particular locus is very variable between individuals and is inherited reliably. Over evolutionary time periods, the frequency of changes (mutations) in copy number at such loci is quite high and so there is a lot of genetic variation in the human population. An individual might, for example, have a paternal chromosome carrying 10 copies of a 4 nucleotide STR at a particular locus and a corresponding maternal chromosome with 12 copies. This means that his DNA sample will contain equal amounts of sequences of 40 (4×10) and 48 (4×12) nucleotides in length. The 40- and 48-length variants are both inherited in a simple Mendelian manner. Other individuals might have different length variants at the same locus, for example 8 and 9, or 13 and 16, or 2 copies of 10. The measurement of the length of the STRs at this particular locus in the genome is done by a technique known as the polymerase chain reaction (PCR). Here short synthetic sequences of DNA complementary to the non-repeated DNA on either side of the locus are prepared by chemical synthesis. These can initiate DNA replication by a DNA polymerase enzyme plus the four deoxynucleoside triphosphates. The PCR reaction involves repeated cycles of DNA synthesis, strand separation and reattachment of the synthetic primers. The end result is the synthesis of a large number of copies of the DNA at the locus of interest. The reaction products are separated by a suitable technique, such as capillary electrophoresis, and the molecular sizes of the products obtained indicate the number of repeats at this locus in this DNA sample.

If just one locus is examined, many individuals will share the same markers, but if several loci are looked at, the probability of a random match for all of them between any two individuals falls until it becomes millions to one against. The technology of DNA-based forensic identification changed rapidly every few years when it was first introduced, but in the 2000s it became standardized on the basis of a certain number of loci containing different STRs. In the USA, databases are compiled using a set of 13 loci. In the UK it is ten, in Germany eight. A standard set of seven loci is used for comparing results between different European databases. In addition to the STRs, an actual gene is also included in the standard test to determine sex. This is a gene encoding the protein called amelogenin, which is concerned with tooth enamel production. It occurs on both X and Y chromosomes as different length variants and therefore measurement of its length allows the identification of sex. The UK National DNA Database is currently the largest in the world containing data on about six million individuals in 2012. It is the largest because in the UK DNA samples are routinely collected from all people arrested for any reason, something not usually done in other countries. Although samples from innocent people are now supposed to be removed from the database, this has been slow in implementation. Such a large database does raise issues about privacy and civil liberties, but so far the UK population appears to have acquiesced to the measure on the basis of its effectiveness in detecting crime. When crime scene samples are screened against this database, there is about a 50 per cent chance of finding a match.

The forensic use of STRs may seem a very efficient method of detection. However there are various reasons why it would be unwise to convict anyone on DNA evidence alone. First, samples can often be small, damaged, and partially degraded. This means that the amplification reactions do not work properly and some markers can be missed. Second, samples may contain DNA from more than one individual, giving mixed profiles. Third, the very low probabilities of a match usually quoted are based on the assumption that members of the population are unrelated to each other. Obviously relatives are likely to share some of the markers, so for a given subset of people the probabilities are not always as astronomically small as they seem at first sight. Fourth, the bigger the database the greater the probability of a completely chance match. The consequences for a completely uninvolved individual accused of a serious crime on the basis of a DNA match, accompanied by no other evidence, would be most unpleasant. Such a person would certainly be under great pressure to prove their innocence in the face of the apparently millions-to-one detection specificity.

On the positive side there are numerous examples of the success of DNA fingerprinting in helping to convict, or to exonerate, suspects in many high profile criminal cases. It is also possible to do retrospective analysis so long as suitable samples are available. One well-known case is that of James Hanrattay, one of the last individuals to be legally executed in the UK. Hanrattay was convicted for the murder of Michael Gregsten after hijacking his car and forcing Gregsten to drive to the A6 road (hence the 'A6 murder'). Also in the car was Gregsten's mistress, Valerie Storie, who was raped by Hanrattay, and shot four times resulting in her subsequent paralysis. Hanrattay was convicted and hanged in 1962. However this was a complex case and it generated considerable controversy for many years with high profile advocates continuing to claim his innocence. In 2002 his body was exhumed and DNA samples taken. His DNA turned out to be a match for samples found on Storie's underwear and on a handkerchief in which the murder weapon was wrapped. No other DNA was found. The Appeal Court judges considered on this basis that Hanrattay's guilt was confirmed 'beyond doubt'.

The same method used for criminal investigation is also used for other types of identification. One common application is the establishment of paternity. A baby will have about 50 per cent of markers from the mother and 50 per cent from the father, so if the putative father has a match to one of the markers at every locus the probability that he is the real father is extremely high. Exactly the same argument is used to resolve immigration disputes about whether an individual is or is not a first degree relative of a lawful resident, which may mean they are also entitled to residency. Another application is for victims of disasters, for example air crashes or mass killings, where there may be many bodies burned beyond recognition. This requires samples from family members, or an authentic pre-disaster sample from the individual. Identification is based on the principle that first degree relatives (parents, children, siblings) will share about 50 per cent of markers, second degree will share about 25 per cent, and so on. With all the applications involving comparison of relatives it should be borne in mind that new mutations do sometimes occur and will mean that individuals do not match the parents with respect to both markers at the mutated locus. Furthermore mutations occurring in non-reproductive cells, especially those occurring early in embryonic development which are later present in a large fraction of the body, means that one individual can generate an apparently mixed sample.

Ancestry and migration

In recent years, gene variants have been extensively examined to assess human ancestry and migration. Particular use has been made of neutral variants on the Y chromosome, which are specific to the male lineage, and those on mitochondrial DNA, which are specific to the female lineage. This is an area where misunderstanding is common and to avoid this it is important to understand how inheritance of these markers differs from those in the rest of the genome.

Mitochondrial DNA

Mitochondria are small bodies found in all types of cell and are responsible for the generation of energy by the controlled oxidation of the products of metabolism. Although most of the proteins that make up mitochondria are encoded by nuclear genes, mitochondria also contain their own DNA, abbreviated as mtDNA, and this encodes some of the proteins and RNA molecules making up the structure. Mitochondria can grow and divide along with the cells that contain them, and their DNA can be replicated, so each time mitochondria divide, some mitochondrial DNA copies are distributed to each new mitochondrion. Because of these facts, and the similarity of the mtDNA to that of bacteria, it is thought that mitochondria were originally bacteria in the remote evolutionary past (see Figure 8). Usually, all the mitochondria in the body have the same DNA sequence, although it is not uncommon for two or more sequences to be present at measurable abundance. Because each mitochondrion contains several DNA molecules, and each cell contains many mitochondria, the overall abundance of any particular locus in mitochondrial DNA is much higher than in nuclear DNA where there are only two copies of each gene. Hence it is easier to recover useful quantities of mtDNA than of nuclear DNA from degraded samples, for example archaeological specimens.

Unlike nuclear DNA, which is inherited equally from both parents, mtDNA is inherited only from the mother. This is because the mitochondria of the sperm are destroyed shortly after fertilization and only the mitochondria of the egg survive. The mutation rate of mitochondrial DNA is much higher than that of chromosomal

DNA, mostly because the DNA repair systems of mitochondria are poor. If a mutation occurs, it will most likely be lost due to random sampling of mitochondria in each cell division. However there is a small probability that it will spread, also by chance, until it becomes the normal version of that locus in the mitochondrial DNA of the whole body. This process occurs mostly during the earliest stages of female germ cell development, when the future mother is herself an embryo. At this stage there are only a few mitochondria in each of the newly formed reproductive cells in the embryo and so the probability for a new mutation to become the normal variant by chance is relatively large.

These properties: maternal inheritance, high copy number, and high mutation rate have meant that mtDNA has been extensively used for ancestry tracing. The technology used for these studies has evolved rapidly. Nowadays it is common to determine the sequence of the entire mitochondrial genome, which is quite small (16,569 base pairs), although many studies have been done in the past in which only a hypervariable non-coding region is sequenced.

If a large number of DNA sequences from different individuals are compared, some will be more similar than others and a 'family tree' can be constructed for the whole set of sequences. There are several different algorithms using different computational methods for making these trees but in general the more similar sequences are considered more closely related. Each sequence can be identified in terms of the specific set of markers it carries at each of the variable positions. This is called the 'haplotype'. A set of haplotypes arising from one branch of the tree forms a 'haplogroup' (Figure 13). These are nested such that the haplogroups that originate from branches near the top of the tree contain those that originate lower down. For example, in Figure 13, mutation number 1 occurs in the ancestral sequence and the complete set of sequences containing mutation 1 is considered to be 'haplogroup 1'. Because mutations numbers 2, 3, and 4 occur

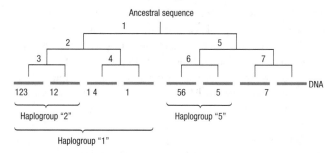

13. Haplogroups. The diagram shows eight DNA sequences derived from an evolutionary history involving seven separate mutations, labelled 1–7. Once a mutation has occurred, it persists. Three of the resulting haplogroups are identified: '1', '2', and '5' representing the set of sequences derived from the common ancestor that possessed each of these mutations

later on in the lineage, haplogroups can also be defined by them, each of which is a subset of haplogroup 1. For DNA of the chromosomes (other than the Y chromosome), haplotypes do not persist for long in evolutionary time because of the chromosome breakage and rejoining that occurs during meiosis. However for mtDNA, where such recombination is not possible, the haplotypes do tend to stay intact over long periods. The timescale of such a diagram can be estimated roughly from estimates of the mutation rate, because according to a well-known theorem of evolutionary biology the rate of neutral mutations becoming the normal form of the gene is the same as the rate of *de novo* mutation. Mutation rates can be measured directly by comparing mtDNA taken from parents and offspring. They can also be estimated by sequencing mtDNA from archaeological specimens, which, if they are not too old, can be dated because of the radioactive decay of the Carbon-14 that they contained when alive. For mitochondrial DNA such a family tree can be traced back to a 'last common ancestor of all humankind' with a date of approximately 150,000–200,000 years ago. This last common ancestor was, of course, a woman, because mtDNA is inherited only maternally. Although we know nothing else about her, this individual has, understandably, been called 'Eve'. The geographical distribution of the present day mtDNA haplogroups indicates that she must have lived in Africa, sometime between 150,000 and 200,000 years ago. Of course there were many other men and women living at the same time, but none of the others have transmitted their mtDNA to present day humans.

Y chromosome

For all mammals, sex determination occurs through the XY chromosome system. Females have two X chromosomes, and males have one X and one Y chromosome. The other chromosomes are known as autosomes. The X chromosome is much bigger than the Y and it carries many genes, which encode proteins having a whole variety of functions, most of which are not concerned with sexual development. In females one of the X chromosomes in each cell becomes inactivated early in development so that the overall 'dosage' of X chromosome encoded products is balanced with that of the autosomes in the same way as it is in males. The Y chromosome carries a few genes not found on the X chromosome, in particular a gene called *SRY* which is important for determining maleness during embryonic development. Most of the Y chromosome is different in DNA sequence from the X and because of this it does not exchange gene variants by recombination during meiosis. This means that like mtDNA, and unlike autosomal DNA, a set of markers established originally by new mutations, a haplotype, tends to remains together over many generations. The type of variation most often used for ancestry studies is a set of simple tandem repeats found on the Y chromosome. These are analysed in a similar way to the autosomal STRs used for forensic identification which were described above. There are also many single nucleotide changes (SNPs) which are used for analysis.

As for mtDNA, it is possible to make a rough estimate of the time of the last common ancestor of all humans by looking at the family tree of haplotypes in human Y chromosomes. Current estimates are that he lived about 130,000–150,000 years ago. Again, we know nothing about this individual except that he was male, because he possessed a Y chromosome. Not surprisingly, he has been called 'Adam'. There were many other men also living at the same time as Adam, who had different markers on their Y chromosomes, but their male lineage descendants have not survived to the present day. Although estimates vary, the majority of studies have indicated that the time back to Adam is somewhat less than the time back to Eve. This may reflect the propensity of a few men to leave a lot of descendants, a process that accelerates the process of convergence of the Y chromosome tree, as shown in Figure 14. Using the names 'Adam' and 'Eve' for last common ancestors deduced from family trees of Y chromosomes and mtDNA tends to suggest that the Bible story is true and that one couple spawned the whole of humanity. But this is not so. It is probable that 'Eve' and 'Adam' were separated from each other by some geographical distance and many thousands of years, and there were many other people around at the same time as either of them. It is simply an inevitable consequence of mutation and inheritance that there will be a last common ancestor at some point in the past. These ancestors are different people because they are only ancestors in respect of the Y (all-male) and mtDNA (all-female) lines of descent. We all contain many other gene variants on our autosomes which were not possessed by Adam or Eve and which descended through different lineages.

The relative ease of conducting studies using mtDNA or the Y chromosome have led to the emergence of a number of companies who can test your DNA and 'tell you your origins'. Unfortunately the results of such studies are often misunderstood by the clients. For example, if an African American has such a test he may be told that he comes from some particular tribe of west Africa, from an area where the slave population of America was originally taken. He may feel an affinity with this tribe and feel he has

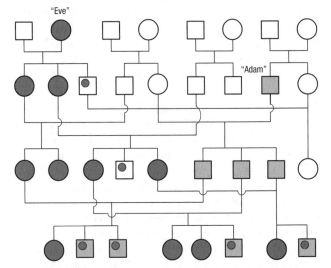

14. **Common ancestors.** This diagram illustrates why it is possible for the whole population to be perceived as descending from 'Adam and Eve', based on Y chromosome and mitochondrial DNA. However, these two were not the only people in the world, they did not necessarily meet or live at the same time, and they are not ancestors of the current population in terms of all the autosomal genes. Circles indicate females and squares, males. Dark shading represents Eve's mitochondrial DNA and light shading represents Adam's Y chromosome

discovered his 'roots'. However, it is only one root. Because mtDNA is inherited in the female line and the Y chromosome in the male line they do not tell you very much about your ancestry. If any individual man, for example, were to contemplate the previous 14 generations, he may have as many as 16,384 direct ancestors. This is the maximum number assuming that all generations are outbred, as any consanguinity would reduce the total. The results from the company will probably only tell him about one or a few ancestors. Using those autosomal markers that show a significant difference of frequency between European and African, it has been estimated that the black population of the

USA have 10–20 per cent of 'white' admixture in their genomes, reminding us that people's true roots are always complex.

One interesting cluster of related variants on the Y chromosome occurs in central Asia, comprising 0.5 per cent of the whole world's population. This cluster has a calculated time to its last common ancestor of about 1,000 years. It has been suggested that the ancestor in question was Genghis Khan, founder of the Mongol empire. Genghis had a very large harem, and so did his sons, so the potential for a massive spread of his Y chromosome markers is certainly plausible.

Geographical history of humanity

Notwithstanding the above caveats, it is possible to make some deductions about the major past migrations of human populations by examining the geographical distribution of genetic markers. Unfortunately the picture has tended to become less clear the more data have been collected and analysed. It is known from archaeological studies that recognizably human species appeared first in Africa. About 1.8 million years ago the species called *Homo erectus* migrated across Europe and Asia. *Homo erectus* may or may not have given rise to other non-modern species including *Homo neanderthalis* (Neanderthal man) and *Homo floresiensis* (The 'Hobbits' from the island of Flores in Indonesia). But some of these possible descendant types persisted until relatively recently and certainly within the time frame of modern human occupation. The main issue has been whether modern humanity arose by independent evolution of local races from an earlier species such as *Homo erectus*, or whether a new population of modern humans swept across the globe and replaced all the other species who were already living there. In general the genetic evidence strongly supports the latter hypothesis, sometimes called the 'Recent African Origin' or 'Out of Africa' theory, although there are odd bits of data suggesting some admixture of genes from more ancient sources.

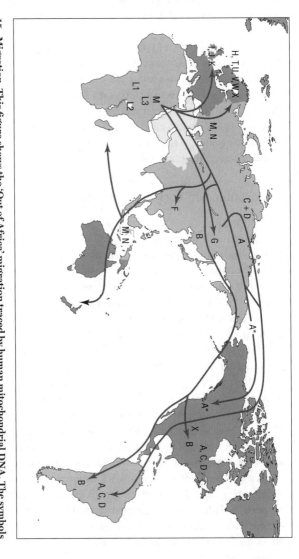

15. Migration. This figure shows the 'Out of Africa' migration traced by human mitochondrial DNA. The symbols indicate haplogroups. The African haplogroup L3 generated M and N which were possessed by the migrants from Africa. Various haplogroups arose from N in Europe and a different set of haplogroups from N and M in Asia. Native Americans have a subset of the Asian haplogroups, indicating an Asian ancestry

The issue has been studied using data from mtDNA, Y chromosomes, and also markers from autosomes. The principle is to construct family trees from the sequences, define haplogroups from the branch points (one haplogroup contains all the descendants from a particular branch point) and then look at their geographical distribution. If the interest is in ancient events, then known modern migrations such as the European settlement of America or the Atlantic slave trade are discounted. The distribution is consistent with the idea that modern humans arose in Africa, migrated out of Africa about 60,000–90,000 years ago, and sequentially populated Eurasia, Oceania, and the Americas (Figure 15). The reasons for believing this are that the most ancient haplogroups are all found in Africa and some of them are confined to Africa. So in effect the rest of the world's population is a subset of that found in Africa. The time of the migrations from Africa (which may actually have been many separate events) can be estimated by calculating the time of the branch points representing the sub-haplogroups not found in Africa. In general the times of occupation of the various continents are broadly consistent with archaeological evidence.

Race

Given that all humanity arose from a migration, or set of migrations, that was relatively recent in evolutionary time it is pertinent to ask whether distinct human races really exist. They certainly exist in people's perception, as people define themselves as belonging to specific cultural, ethnic, or linguistic groups. But is there any genetic basis for race? Because of the relatively recent origin of *Homo sapiens* the total level of genetic variation is quite low compared to that found in other species. Moreover numerous studies using all possible methods of examining genetic variation concur that the 'between group' variation is only about 5–10 per cent of total variation. In other words two white Germans will, on average, have nearly as much genetic difference between them as either one has to an African or Native American.

So, if by race we understand distinct subspecies with a significant genetic difference between groups, then there are no human races. There is nonetheless some genetic difference between human populations. Obviously those gene variants responsible for dark skin pigmentation (still poorly understood biologically) are much more prevalent among Africans than among northern Europeans. Despite the fact that only a small minority of variation is between groups, it is nonetheless possible statistically to reconstruct some racial groupings from large sets of data on human genetic variation. If the number of groups fed into the analysis is predefined as five, then the groups obtained correspond to fairly conventionally defined races: African, Eurasian, East Asian, Oceanian, and Native American, although it should be noted that the Eurasian group in this analysis is quite wide, including Europe, the Middle East, central Asia, and India. If the number is predefined to be larger than five then the groups that emerge become less familiar, for example at six the Eurasian group subdivides into the Kalash (a people from northwest Pakistan) and the rest. So there is indeed some genetic variation correlating with conventionally defined race, but it must be borne in mind that the proportion of genetic variation correlating with these groups is small. This means, for example, when considering racial propensity to develop particular diseases, that there will be some differences, such as the relatively common occurrence of Tay-Sachs disease among Ashkenazi Jews mentioned in Chapter 3. But most hereditary susceptibility to most conditions will vary on an individual rather than a racial basis. Likewise, there has been much discussion about using the forensic typing of DNA to deduce the race of a suspect. But this is not possible from the forensic markers currently in use, as they are found in all races. It is actually very difficult to find variants that are specific to conventionally defined races and at the same time common enough to be of any use.

To summarize, genetic analysis has shown that the 19th century view that humanity consisted of several subspecies which have evolved independently in their current locations is completely wrong. Modern humans are all descendants of the migrations from Africa that occurred about 60,000–90,000 years ago. Most genetic variation is between individuals, but there are also some variants whose frequency differs systematically between groups.

Chapter 5
Genes of small effect

So far we have considered the properties and importance of genes and markers whose deoxyribonucleic acid (DNA) sequences are known. Here we take a jump away from DNA and consider unidentified genes whose variants collectively affect some characteristic of interest. This is the domain of quantitative genetics, a branch of the science established in the early years of the 20th century. The idea is that many aspects of living organisms depend not on the action of a few genes but on the actions of many, each having a small effect on the overall characteristic. This assumption has been used successfully to inform the breeding of agricultural animals and plants. But some of the concepts have also been very controversial when applied to human beings.

Quantitative genetics deals with continuous characters, such as height or weight, which can adopt any value and where animals or plants are usually distributed across a continuum of values appropriate to their species. It also deals with threshold-based characters where the character itself is either present or absent, but its presence is presumed to depend on a continuous susceptibility which depends on many gene loci, such that the trait occurs above a particular threshold level of the susceptibility. The genes whose variants are responsible for quantitative variation are called 'quantitative trait loci' (QTLs). These are not necessarily a special kind of gene. Their status as QTLs exists only in relation to a particular quantitative character. The magnitude of their effect depends on the variants actually present in the population and they may also have larger effects on other traits not under study. It has been difficult in the past to identify QTLs with specific genes although this has become easier thanks to modern whole genome approaches to the problem.

Heritability

When it comes to human beings, people are obviously very diverse in their appearance and characteristics. It is often asked how much of this difference is due to differences of gene variants between them, and how much to differences in environment, in other words differences of nutrition or upbringing or life experiences. Statisticians speak of 'variance' which is a way of expressing quantitative variation in a mathematically tractable manner. The *heritability* of a character is defined as the proportion of the variance in a population attributable to genetic variation, with the balance presumed to arise from environmental effects. So if variation in a character is entirely due to genetic variation the heritability is 100 per cent, and if it is entirely due to environmental variation the heritability is 0 per cent.

Heritability concerns *variation* and should not be confused with genetic determination of a character. For example humans have two legs and development of two legs is entirely under genetic control. However the variation in leg number among the human population is almost entirely environmental, due to accidents or military engagements leading to loss of legs. So the heritability of human leg number is very low, notwithstanding the obvious role of genes in the development of legs.

There are many examples of 100 per cent environmental determination, for example the language spoken by an individual, or the religion he or she believes in, depends entirely on the circumstances of upbringing. The types of single-gene genetic disease considered earlier are effectively 100 per cent genetic, but whether any quantitative character is 100 per cent genetic is not known. Heritability in human genetics is often estimated from twin studies. Twins may be identical or non-identical. Identical twins are called 'monozygotic' (MZ) meaning that they arise from division of a single embryo at an early stage of development. Non-identical twins are also called 'dizygotic' (DZ) to indicate that they come from two different embryos. In terms of genetic variation, MZ twins should be completely identical while DZ twins will have the same relationship as other siblings, sharing about 50 per cent of variants. The calculations can be quite complex but in essence the heritability of a particular character is deduced from the comparison of the variation seen between pairs of MZ and DZ twins.

There are a few issues to note at this stage. First, the calculation of heritability from twin data gives something called 'broad heritability'. This includes all genetic effects including dominance and interactions between genes. There is another type of heritability called 'narrow heritability' which deals just with gene variants showing additive effects. This measure is needed for calculating response to selection so is relevant to animal and plant breeding studies.

Second, in the human arena twin studies have been controversial for many reasons. One is that twins raised in the same family will have very similar environments and this will make their heritability seem larger than it really is for the whole population. This can be overcome by examining MZ twins separated shortly after birth and raised apart. But it is hard to find such examples, and many separated twins are actually raised by other family members so their environments are still quite similar. In addition the establishment of MZ versus DZ status can be uncertain if the individuals themselves are not available for examination.

In general it is likely that heritabilities are overestimated by twin studies. On the other hand, in recent decades much more sophisticated statistical methods have been used to analyse twin data than were used in the past, and some of the problems have been overcome by these superior methods of analysis.

Regardless of the actual values measured, a key thing to understand about heritability is that it deals with variation, not genetic determinism, and that it is entirely context-specific. The level of heritability depends on how much genetic variation affecting that character is present in the population under study, and on what is the range of environmental variation to which members of this population are exposed. Consider a population of cattle kept on several farms of different quality. The heritability of growth rate or final size might be 50 per cent. Then suppose one farmer buys up all the other farms and institutes a common policy of nutrition and husbandry. This means that the heritability of growth rate will go up. Although nothing has changed with regard to the genetic variation, the heritability goes up because the environmental variation has been reduced. In fact the heritability of milk yield of dairy cattle in the USA since 1970 did increase from 25 per cent to about 40 per cent, for this type of reason. Conversely, suppose that the cattle are bred for many generations with selection for an optimum growth rate. After several generations the amount of genetic variation falls because several of the key variants have either been lost from the population or have become the normal version of the gene present in all the animals. If the environmental variation is kept constant the heritability of growth rate will fall because the amount of genetic variation in the population has been reduced. This context-dependent nature of heritability is often not understood, but it means that a measurement is only valid for a specific population in specific circumstances.

Many scholars have also been uncomfortable about partitioning human characteristics between 'genetic' and 'environmental' and felt that there should be some greater subtlety to the situation. Undoubtedly there will be situations in which variation of genes and environment interact. For example a particular gene variant might make it easier for an individual to find food in a variegated environment but have no effect in a uniform environment. Interaction of this sort will tend to increase the estimate of heritability based on twin studies because the logic of the study is based on looking at individuals with different degrees of relatedness rather than at individuals of the same genotype exposed to different environments.

Less discussed is the fact that there are other potential causes of differences between individuals which are neither due to gene variants nor to environmental effects. These lie in the domain of random events occurring during development. One such event is the movement of mobile DNA elements called transposons from one genetic locus to another. There are numerous virus-like elements present in the human genome. Most of them are defective and simply get replicated each generation along with the rest of the DNA. But a few still retain the capacity to excise themselves from their location in the DNA and to insert themselves elsewhere. This occurs rarely on a per cell basis but much more commonly in developing neurons than in other cell types. Like any other mutation, the insertion of a transposon at a new site in the DNA may affect the properties of the cell. Whether this is a significant cause of individual variation in personality or mental ability is not yet known, but it is a real possibility.

Genome-wide association studies (GWAS)

Many common human diseases have significant heritability. For example type 2 diabetes, schizophrenia, or heart disease come into this category. One of the goals of the human genome project was to identify gene variants present in the human population which predisposed to conditions such as these. This would have three important benefits. First, it would identify genes involved in the pathology of these diseases and thereby inform the development of new drugs targeted against the relevant gene products. Second, it would enable the prediction of risk for individuals at an early stage in their lives. People could be screened for high risk markers and, when identified, could be given preventive treatment. Third, it would potentially allow for more discriminating treatment as the various subtypes of a disease which may depend on different gene loci could be identified. Most drugs have variable effects on different people and so the overall targeting and effectiveness of drugs could theoretically be improved by understanding how they operate in the presence of the various relevant gene variants. The second and third of these potential benefits are often referred to as 'personalized medicine'.

To achieve all these goals it is necessary to know the genomes of very many people. The huge improvements of DNA sequencing technology and the reduction of cost of sequencing to a few thousand dollars per genome have enabled many studies to be carried out in recent years. Many of these are 'genome-wide association studies' (GWAS). These do not involve total genome sequencing but instead map very large numbers of markers involving single nucleotide changes (called single nucleotide polymorphisms, or SNPs) to find which markers correlate with particular conditions or diseases. Usually the SNPs themselves are not in the genes whose functions are being detected, but are presumed to lie fairly close to them. In a population reproducing at random for an infinite time, the markers present at all loci would be randomized relative to their neighbours by the crossing over between parental chromosomes that occurs at meiosis. However in a real population this process occurs so slowly that there usually remains some association between markers that lie close to each other in the DNA. This is the basis of the association method. About 500,000 SNPs provide sufficient resolution to do these studies in humans. In a GWAS the frequency of each SNP type is examined in individuals with the condition under study and those without. If there is a significant difference, then the region of the genome containing this SNP marker is presumed also to contain a functional DNA sequence variant that affects the condition in question.

Despite the potential power of the technology, it has to be admitted that the results from GWAS to date have been rather disappointing. Some new gene variants affecting each common disease or condition have been found, but these variants tend only to account for a small fraction of the total heritability of the trait deduced from twin studies or other quantitative genetics methods. The reasons for this 'missing heritability' have been much debated. The most likely reasons are that the complex conditions examined are affected to only a small extent by common variants, and also that there are a large number of rare variants at work. Beyond a certain limit, the sample sizes used in GWAS are not sufficient to detect such effects.

Even if larger studies were done, and complete genomic sequencing were done instead of SNP profiling, in practical terms this means that many of the great hopes of the human genome project are unlikely to be realized. It will not be possible to predict who will develop type 2 diabetes, or other common diseases, even from a complete genomic sequence, because the causative variants are too numerous in the genome and each of them is individually too rare in the population. This is in stark contrast to the situation with single-gene disorders where a gene sequence gives a very high level of predictability of the risk of developing conditions such as haemophilia or cystic fibrosis. For similar reasons it is also unlikely that genomic data will be very helpful in understanding pathology or be able to inform a personalized drug regime in such conditions.

Human height

Human height is a familiar characteristic that exemplifies many of the principles of quantitative genetics, where variants of many genes each have small effects on the overall trait. Height is a very familiar attribute (Figure 16). It is easy to measure and many accurate records exist over long time periods, particularly for army recruits. It is generally recognized to have a considerable heritability and because of this, and the shared environment within a family, it is possible to predict the adult height of children fairly accurately, knowing the heights of their parents. Unlike some other issues to be discussed below, height also has the advantage of being non-controversial, so everyone can assess the issues and findings without prior ideological bias.

The biology of the development of height is understood to some extent. Maternal nutrition is important, with malnutrition during pregnancy leading to smaller babies as well as many other long term health problems. Early childhood nutrition is critical because growth rate is very fast just after birth. Final height depends largely on the elongation of the long bones of the legs, which depends on the growth rate of the cartilage at the ends of the bones. This terminates in late puberty with the fusion of the bony caps to the shafts and the change of the remaining cartilage to bone. Among the factors affecting cartilage growth is an inhibitory effect of a group of substances called fibroblast growth factors (FGFs). As discussed in Chapter 3, abnormally increased FGF activity inhibits cartilage growth more than normal and this is why individuals with gain-of-function mutations in the gene for FGF receptor 3 are achondroplastic dwarfs. However this group of mutations does not contribute significantly to the variation of height among normal individuals. This illustrates once again the fact that heritability relates to the variation actually existing in the population, not to the mechanism of the process. It may be that there is little or no genetic variation in the genes controlling the principal mechanism of a process. Instead the genetic component of variation is mediated through other genes that may have other principal functions and only small or incidental effects on the process under study.

16. Tall and short. The extremes of variation of human height

Estimates of the heritability of human height range between 65–90 per cent, depending on the particular population under study. The lower values around 65 per cent relate to studies from China and Africa, where environmental variation in terms of nutrition and health care is greater than in Europe or North America. The higher values relate to individual nations in northern Europe which are relatively homogeneous in environment.

Genome-wide association studies on about 30,000 people have recently been conducted to identify the source of this variation. About 50 SNPs were found to have a significant association with height. Of the genes lying near enough to the SNPs to be candidates for the functional variation, some had previously been identified as involved in some aspect of growth, either through animal experiments or through knowledge of rare human genetic syndromes with more severe mutations in the same genes. However the amount of heritability attributable to these loci is at most 5 per cent of that measured by quantitative genetic methods. So only this much heritability can be considered 'explained'. It is likely that the remainder is distributed across a very large number of other gene variants each of which contributes a very small effect. This means that it is not possible, and probably never will be, to predict a person's height from examination of their genome sequence.

The situation is markedly different in the domestic dog. Dog breeds vary hugely in size, from chihuahuas weighing about 1 kilogram (kg) up to Saint Bernards weighing about 80 kg. Much of the difference in dog size is due to a single variant in the gene encoding a growth factor called insulin-like growth factor 1 (IGF1). This is released by many body tissues in response to growth hormone from the pituitary and it stimulates cell division of many cell types. Dogs have been subject to intense selective breeding for thousands of years and this process has amplified *IGF1* gene variants to a prominent role. In humans, IGF1 has the same functions as it does in dogs, and it is just as important in normal development, but variation in the *IGF1* gene is not a significant cause of the variation of human height.

Although height does have a high heritability and some of the genetic variation responsible for it is now identified, this does not mean that height cannot be altered by the environment. In fact human height has increased considerably during the 20th century. In Holland the average height of mature males increased from 172 to 184 centimetres (cm) over the period 1913–80 and the Dutch are now the tallest people in the world. The height of American adults increased by a somewhat lower amount, from 173 to 179 cm (white males) over the same time period. In the mid-19th century Americans were the tallest people on Earth, but in the 20th they were overtaken by some European populations, particularly the Dutch. The explanation for this is presumably that in the 19th century, American nutrition in pregnancy, and in childhood when the long bones are growing, was, on average, superior to Europe's, and enabled more people to reach their full potential height. In the 20th century European nations caught up in prosperity and consequently in childhood nutrition. Why have they recently surpassed the Americans? One possible reason is that health care in the USA is very unequal, with the uninsured populations receiving minimal antenatal and postnatal care, while in Europe state health systems tend to provide better service to the poor section of the population.

Another possible reason is the presence during the 20th century of large scale immigration into the USA of populations who are intrinsically shorter, notably from Latin America and Asia. This explanation brings up the issue of the difference in average height between populations and to what extent they are genetically determined. Such differences do obviously exist. The average height of an Indian adult male is currently 165 cm, of a Chinese 168 cm, and of a Japanese 171 cm, compared to 179 cm for Americans and Europeans. Asian heights are lower than those of North America and Europe. However the gap has been closing with the rapid economic development of Asia and it is not yet known how large the residual genetic difference will be when the Asian populations have all reached the Western standard of living and medical care. It is possible that there will be no genetic difference at all, but in all likelihood there will turn out to be some difference.

It has been presumed in the past that the between group differences are at least partly genetic, like the within group differences. However this does not logically follow and the same issue has been very contentious in relation to intelligence quotient (IQ; discussed later). Two populations might have an equal heritability for height, but one has a better environment than the other and is therefore taller on average. In this case the reason for the between group difference is 100 per cent environmental, despite the importance of heredity in determining individual height within a group. For example, there is a current 8 cm difference between the height of adult males in North and South Korea. It is unlikely that there is much genetic difference between the two Koreas so this is probably attributable to famine and poverty in the north versus the booming economy in the south. In comparable situations, migrants from Guatemala to the USA during the Guatemalan civil war achieved a height 10 cm greater than those who had stayed in Guatemala. In the 1950s, males of the Dinka people of Sudan, then the world's tallest, measured 182 cm. Now, following decades of civil war and deprivation, the average has fallen to 176 cm.

Serious mental illness

Although the issue has generated considerable controversy there is substantial evidence that the most serious mental conditions have a high heritability.

Schizophrenia is a common disorder with a lifetime risk of about 1 per cent. It involves delusions and hallucinations, disorganized thought, social withdrawal, and abnormal behaviour. Bipolar disorder is a spectrum of mood disorders affecting about 1.5 per cent of people. It is characterized by disturbances ranging from extreme elation to severe depression. It is associated with high morbidity and a 10 per cent suicide risk. Both conditions cause a lot of individual suffering and are costly in terms of long term medical and social care.

For schizophrenia, twin studies indicate a heritability of around 85 per cent. Similar results come from other family correlation studies and adoption studies. Gene variants associated with the disease have been sought by various methods, most recently by GWAS using large numbers of individuals and examining large numbers of SNPs. Some candidate genes have been identified in individual studies but most of these have not been found again in subsequent studies. Those that have been found in several studies, and therefore seem well established, confer only a small increased risk of up to 1.4 times the normal level. Some of the genes located near the disease-associated SNPs have a likely involvement in neuronal function or development, for example they encode cell surface or structural molecules important in neurons, while others do not. Estimates of heritability for bipolar disorder are also about 85 per cent. GWAS of bipolar disorder have given similar results: no common variants show high association but some are associated with modest risk increases. Interestingly some of the chromosome regions associated with schizophrenia and with bipolar disorder turn out to be the same, suggesting a possible connection between the mechanism and pathology of these psychoses that was not suspected before.

Recently a strong association has been found between schizophrenia and large structural variations of chromosomes (deletions and duplications covering more than 1,000 base pairs). These are collectively quite common but individually rare, meaning that different individuals are likely to have different variants. Such structural changes may affect many genes and they may exert their effect in various ways, for example by altering the number of copies of a gene, such that the level of activity is too high or low, or by disturbing regulating sequences associated with genes. Structural variations are also associated with a range of other neurodevelopmental disorders including autism, some types of learning disability, attention deficit hyperactivity disorder (ADHD), and idiopathic generalized epilepsy.

At present, despite enormous resources devoted to the genetics of serious mental illness, it must be admitted that the genetic basis of the common diseases remains unknown. It seems unlikely that there will turn out to be small numbers of common gene variants causing either schizophrenia or bipolar disorder. More likely there are large numbers of rare variants such that every individual's disease is actually a different disease at the genetic level.

An example of a neurological condition with psychiatric symptoms that does have some degree of genetic understanding is Alzheimer's disease. This is very common, accounting for about 2/3 of dementias. Initially, the usual indication is difficulty with short-term memory. As the disease progresses, additional symptoms may include aggression, irritability, mood swings, language problems, and loss of long-term memories. This is followed by a withdrawal from all social interactions and a decline in bodily functions, with death as a final outcome. The disease is associated with appearance of plaques in the cerebral cortex containing a protein fragment called amyloid beta. It is also associated with tangles of a type of protein filament normally found in neurons. The amyloid beta is produced from an amyloid precursor protein (APP), by two enzymes called presenilin 1 and 2 (PS1, -2). The usual form of the disease increases steeply in incidence with age. However individuals carrying mutations in the genes encoding either the APP or one of the presenilins often get the disease in earlier life (early onset familial Alzheimer's disease). These mutations increase the rate of production of amyloid beta. They are gain-of-function and therefore dominant genetically. In addition, individuals carrying a particular variant of the gene for the protein apolipoprotein E (APOE, variant E4) have a three-fold increase of risk of developing the normal disease, and if they have two copies, the relative risk is 15-fold. APOE helps break down the beta amyloid peptide, and the E4 variant is less effective at doing so than the normal form.

Alzheimer's disease differs from schizophrenia or bipolar disease insofar as the pathology is at least partly understood and there are some known genetic variants that either cause the disease or make it more likely. However the *APP*, *PS1*, and *PS2* gene variants are very rare and only account for 0.1 per cent of Alzheimer's disease. The APOE effect has much more impact on disease incidence, but still does not account for all the heritability. The heritablity of Alzheimer's disease has been estimated at 59 per cent from twin studies, less than that measured for the psychoses but still indicative of a high element in the risk due to genetic variation. As with the psychoses, most of this heritability remains unexplained.

IQ

No issue in genetics has been as bitterly controversial as that of the inheritance of intelligence, usually in the form of the number measured by IQ (intelligence quotient) tests. Even more explosive is the associated issue of whether the measured differences in IQ scores between different 'races' have a genetic basis.

Intelligence, as measured by IQ tests, is a measure of mental ability that is intended to be unaffected by culture or education, and to predict performance in tasks or occupations requiring mental prowess of various sorts. Advocates of IQ tests argue that an individual's IQ tends to remain constant from about age 16 through adult life, and that there is a good correlation with life prospects including seniority and responsibility at work and even salary in many occupations. Critics of IQ tests consider that intelligence cannot be reduced to one number and that the tests are always biased in one way or another. Moreover it is argued that they are self-confirming, as a new test needs to be calibrated before use by giving it to a number of people and then adjusting the results so that they form a normal distribution ('bell curve') with a mean of 100 and a standard deviation of 15 or 16. The properties of a normal distribution mean that that about 95 per cent of the population lies within two standard deviations of the mean—between 70 and 130—and about 2.5 per cent lie above and below this range.

There have been many attempts to devise tests of more specific mental abilities and one often used distinction is between 'fluid intelligence' which is the ability to solve novel problems by using reasoning, and 'crystallized intelligence' which is a knowledge-based ability dependent on education and experience (these are known as 'performance' and 'verbal' measures using the Wechsler scale). Fluid intelligence tends to decline with age in adult life, while crystallized intelligence is largely resistant. However, the principal debate focuses on the general type of IQ test intended to measure a generalized mental ability, sometimes called 'g'.

In principle it seems certain that gene variants ought to affect mental ability, as they affect all other aspects of the organism. Although human mental creativity can be extremely subtle and complex, gene variants affecting intelligence could exert their effects through relatively simple means, for example by affecting the propensity of neurons to form connections with other neurons. There are many individual gene variants known in humans that cause mental retardation due to errors in brain development, so presumably there must be variants that have lesser effects on mental abilities lying within the normal range. There are no well-established genes that cause high intelligence in humans but a few have been found in animal experiments. For example, overexpression of the gene encoding a receptor for one of the neurotransmitters in the brain (N-methyl-D-aspartate) has been reported to increase the maze-solving abilities of mice.

Evidence that intelligence, as measured by IQ tests, has high heritability, is derived from the usual sources of twin studies, other family comparisons, and adoption studies. The field got off to a bad start as the early work, conducted in Britain by Sir Cyril Burt, is now known to be unreliable and parts of it have been considered fraudulent. However there have been many more recent studies, which are more sensitive to the various potential problems, such as whether twins are really MZ, and whether they were really separated, or just being raised by nearby family members. For example, five studies of MZ twins raised apart have estimated heritabilities of IQ between 68 per cent and 78 per cent. On the other hand, studies of children adopted by high IQ families, compared to similar children who were not adopted, show little difference. So there is a prima facie case, based on these traditional quantitative genetic techniques, that the heritability of IQ is high. Molecular genetics has not contributed very much to this debate. Some GWAS have been carried out to try to identify significant variants but so far the results are similar to those of other quantitative characteristics. The associations with individual SNPs reported in one study have not been sustained in others, and it is likely that the genetic variation is distributed across a large number of variants of low effect. A recent study of 126,000 people found that the top ten SNPs accounted only for 0.2 per cent of the variance in years of education achieved, a quantity highly correlated with IQ.

Although the evidence for a significant heritable element of IQ seems good, there are some significant problems. One trend of thought in quantitative genetics that has been evident since the 19th century, although it tends to be less aggressively advanced today, is the theory of racial degeneration. It goes like this. Differences of intelligence are largely inherited. People of lower intelligence (normally of lower social class) have, on average, more children than those of higher intelligence (normally of higher social class). Therefore the average IQ of each generation will be somewhat lower than the previous one and the human race is inevitably degenerating. Such analysis invites a public policy response. The word 'eugenics' was originally coined by Francis Galton, a 19th century English polymath who, among many other activities, was a pioneer in the development of statistics. The term has come to refer to a social policy of improving the human population by selective breeding for positive traits or selective prevention of breeding for unwanted traits. In the 1920s and 1930s there were serious attempts to reduce breeding of less intelligent individuals, especially in Nazi Germany but also in several Western democracies. In the USA several states introduced sterilization laws directed against the mentally retarded and the mentally ill, and sometimes also those suffering from deafness, blindness, epilepsy, and various congenital handicaps. Altogether about 65,000 people in 33 states were sterilized under these programmes, which continued until the 1970s. Eugenics was embraced by many influential and otherwise progressive thinkers in the early 20th century, but since the Second World War it has been seen as cruel or, at the very least, impractical.

The theory of racial degeneration sounds plausible, but the actual evidence points the other way. IQ scores have, in fact, been steadily rising through the 20th century in all countries where measurements have been made. This is known as the 'Flynn effect' after its discoverer. The effect is not immediately obvious because of the convention of calibrating new IQ tests so that they have a mean of 100 and a standard deviation of 15. However, when people are asked to take old tests they perform better than they do on new ones, indicating that the tests themselves have become harder. The Flynn effect is quite substantial, comprising about three IQ points per decade in the USA, and comparable figures in other countries. How long this has been going on is not known, although it seems unlikely that it extends back much beyond the 20th century, otherwise peoples of the past would have had quite remarkably low IQs by modern standards.

The reasons for the Flynn effect are not known, but they cannot be genetic as such large changes would require substantial selection, which has not, despite the wishes of the eugenicists, taken place during this period. There is some evidence that the bulk of the effect is at the lower end of the distribution, consistent with it being due to the various improvements in living conditions during the 20th century. It is interesting to compare the Flynn effect to the concurrent increase of human height, also a rapid and worldwide phenomenon associated with increased prosperity. Presumably the rise in IQ is due to some combination of better nutrition, less infectious disease, better schooling, and more general mental stimulation from newspapers, television, mobile phones, and all the other paraphernalia of modern life. In the recent past the level of IQ may have stopped increasing in the most privileged parts of the world. Studies on British populations since 1980 and those in Scandinavia since 1990 indicate a reduced or zero increase. Those who adhere to the eugenic outlook argue that the Flynn effect represents an improvement of the part of the IQ that is influenced by the environment and has masked a concurrent, but small, degeneration. This is, of course, possible, but the fact remains that IQ has risen and not fallen, contrary to the predictions of eugenicists.

Group differences in IQ

The issue of group differences in IQ has been especially controversial. The debate has focused most on the difference of IQ of about 15 points (about one standard deviation) between black and white populations in the USA. This difference has been measured many times and there is no doubt about its existence. However, there is considerable uncertainty about the reason for the difference.

Logically speaking there is no reason why a difference between two groups need bear any relationship at all to the heritability of the trait within the groups. The well-used example of geneticist Richard Lewontin illustrates this point. Imagine two fields of the same genetically heterogenous corn, one planted in a very fertile soil and the other on poor soil. When the two crops have grown up there will be a large difference of size and yield between them. Although the seed is heterogeneous and so there is a significant heritability for size between different plants within a population, the role of heredity in the difference between the two fields is zero, as this is entirely due to the environment. It has been argued that the same applies to the black–white difference in the USA, especially considering that many socially disadvantaged groups around the world also score lower in IQ tests than their local majority population (these include the Maori (native New Zealanders), the scheduled castes ('untouchables') in India, the Sephardic Jews in Israel, and the Burakumin ('hamlet people') in Japan). The Flynn effect indicates that environmental change can bring about IQ changes of similar magnitude to the black–white difference. Since the civil rights legislation in the 1960s, and the consequent improvement in educational and other opportunities for black Americans, there has been some closing of the gap between black and white in the USA, but there is disagreement about how much closure has occurred and there certainly is still a significant gap persisting to the present day.

Even though most genetic variation is between individuals and not between conventionally defined races, it is nonetheless possible for 'races' to differ systematically in variants that affect a particular quantitative trait. For example this is certainly the case for some disease susceptibilities. It might conceivably also be the case for height or for IQ. But given the large international differences in both these traits, and the speed with which population means have changed in the recent past, it is really not possible to be sure which, if any, of the measured group differences are indeed based on genetic variation.

Chapter 6
Genes in evolution

In the early 20th century NeoDarwinism became generally
accepted among scientists, although even today there seems to be
some way to go before it is accepted by the general public in
certain parts of the world. NeoDarwinism is the synthesis of
Darwin's theory of natural selection with Mendel's genetics, and
the mathematical theory of the process was developed in the
1930s by Ronald Fisher, Sewall Wright, and J.B.S. Haldane. The
mechanism of evolutionary change is considered to arise from
mutations which confer reproductive advantage on the individuals
carrying them such that, over a number of generations, each
advantageous mutation spreads through the population and
eventually becomes the normal version of the gene. A large
number of small changes build up and this may ultimately result
in the formation of a new species, distinct from the original.
Changes arising from natural selection are called adaptations, and
it is adaptation which gives living organisms the appearance of
having been 'designed'. For example the fact that an insect's
proboscis is just the right length to collect nectar from its
favourite flower, or that just the right food is available for an
animal to eat in its normal habitat. Much of the religious
opposition to the theory of natural selection arises from the
fact that it detracts from the 'argument from design' for the
existence of God.

Neutral evolution

However, natural selection is not the only mechanism of evolution. In the second half of the 20th century, studies in molecular biology made it clear that a great deal of change in the primary sequence of deoxyribonucleic acid (DNA) was not adaptive at all. Instead it was 'neutral evolution', consisting of an accumulation of mutations of no selective consequence which spread through the population by the effects of random sampling of variants from one generation to the next. The way this works can be explained by considering a hypothetical very small population, consisting of just one male and one female, reproducing by brother–sister matings. Imagine that a mutation has occurred in one parent, such that half of the sperm or eggs of that individual carry the mutant gene. The chance that a given offspring of the couple carries the mutation will then be 50 per cent. If they have two offspring, the chance that both carry the mutation is 25 per cent. Suppose this occurs. In the next generation, both individuals have half of their reproductive cells carrying the mutation. So there is a 25 per cent chance that a given one of their offspring will carry two copies of the mutation. If they have two children there is a 6.25 per cent (1/16) chance that both will have both copies of the mutation. But if this occurs, it means that the mutation is now the normal form because the previous normal form has been eliminated (Figure 17). This process is called genetic drift and it occurs in all populations because of the fact that each generation of reproduction represents a sampling of the genomes that were available in the previous generation. Of course in a large population the process of establishment of new gene variants is very slow compared to this example of a single couple, and the vast majority of new mutations are eliminated rather than spreading through the population to become the normal form. Nonetheless over evolutionary time there are plenty of mutations and plenty of generations, and as a result genetic drift results in a large fraction of all the evolutionary change that occurs in genomes.

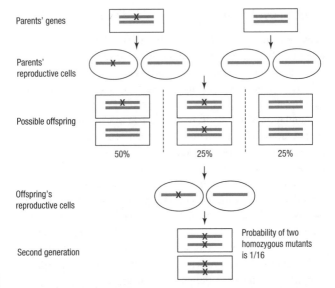

17. **Genetic drift. Neutral evolution can happen by chance. The diagram shows one chromosome pair in a tiny population consisting of one male and one female. Each generation has just two individuals who we must assume are of opposite sex and reproduce by brother–sister matings. One parent has a new mutation, shown as an 'x' on a chromosome. In the first offspring generation there is a 25 per cent chance that both individuals inherit the mutation. If this occurs, then in the second generation there is a 6.25 per cent chance that both individuals have two copies of the mutation**

By definition the type of mutation that becomes established by genetic drift is neutral, meaning that it does not affect the reproductive success of the organism. Because all mutations arise in the context of an existing genome and organism, there is a good chance that any given new mutation will be deleterious. This is because the change it causes is likely to be disruptive to existing mechanisms and so it will probably reduce the reproductive fitness of the organism. In such cases selection does operate in a negative sense and the probability of a significantly deleterious mutation becoming established by chance sampling is effectively zero.

Classification of organisms

Once it was realized that a lot of the variation between species consisted of neutral changes it became clear that this was a new and valuable aid to the classification of animals, plants, and microorganisms. A great deal of modern classification science now uses the primary sequences of genes as the raw data in addition to the traditional anatomical characters.

For any given protein, only a fraction of the amino acids in its structure are necessary for its biochemical activity, so the others are free to change as a result of genetic drift. The DNA in between genes shows even faster neutral evolution than the coding regions of genes, because it has limited function, or no function at all, so the range of base substitutions that can be tolerated without negative selection is increased. Population genetic theory predicts the existence of a 'molecular clock', indicating an approximately linear relationship between the time since the divergence of two lineages in evolution and the number of differences between the sequences of a particular gene.

The result is that the further apart two organism are in evolution, the larger the number of neutral changes in their DNA. If the genomes, or parts of them, are sequenced, then evolutionary trees can be constructed simply from the sequences, without knowing anything else about the organisms. These trees have been very useful in refining areas of classification that were previously obscure, especially for microorganisms, where the amount of anatomical information is often limited, or for large scale relationships, such as that between animals and plants, where there are no common anatomical features (Figure 18).

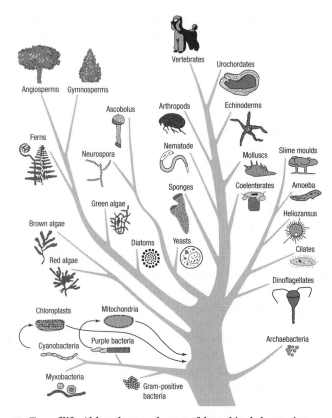

18. **Tree of life. Although somewhat out of date, this phylogenetic tree nicely depicts a wide range of organisms. More accurate trees can now be constructed using the neutral differences between DNA sequences of a set of genes which are present in all organisms**

Although many amino acids at inessential positions in a protein change between species because of neutral mutations, the amino acids at many other inessential positions remain the same. It is the presence of these identities that show that two genes from different organisms are really 'the same' gene. In this context, 'the same' means that they derive from a common ancestral gene and encode a protein with the same biochemical function. The same genes in different species are usually given the same name. For example the cystic fibrosis gene (*CFTR*) exists in all vertebrate animals, although the DNA sequence does differ in base substitutions that do not affect function of the protein, and it is more divergent between a human and shark than between a human and chimpanzee. Many of the genes encoding proteins with basic functions in cell structure and metabolism have remained the same in this sense since the time of the common ancestor of all extant organisms. This is the source of the oft-repeated claim that humans share many of their genes with bananas. What is being referred to here is not gene variants or markers but the actual genes concerned with basic life processes.

Most of the genomic analysis of evolution has focused on neutral evolution because this aspect can easily be explored. But of course it is natural selection that is the more interesting side of evolution because only natural selection can generate real novelty: new structures, new enzymes, and new metabolic processes. Where do new genes come from? To some extent they come from the duplication of old genes. This occasionally occurs due to an error in DNA replication, and it constitutes one type of mutation. If a gene becomes duplicated then only one copy is thereafter required to perform the original function and the other copy is free to accumulate mutations and may eventually assume a different function as a result. Among vertebrate animals, the group which comprises mammals, birds, reptiles, amphibians, and fish, a high proportion of genes belong to multigene families, each family being homologous to just one gene in invertebrate animals. The increase in gene number may have occurred by two duplications of the whole genome about 500 million years ago, followed by the elimination of many of the duplicate copies. It is thought that the opportunities for creation of new functions arising from all these spare genes was an important element in the success and diversification of the vertebrates. Similar duplications may underlie the evolutionary success of the higher plants.

How do we know that genes that are related in sequence are really derived from a common ancestral gene, rather than having evolved by natural selection to a similar sequence? The reason is that convergent evolution driven by functional efficiency would yield proteins that were identical only for the functionally necessary regions. The amino acid sequences in between would have no similarities, so the overall sequences would still be very different. There are some examples known where two genes have converged to the same function starting from totally different origins. In such cases they have no sequence identity at all, or any they do have is confined to a small region of the molecule actually responsible for its biochemical activity. An example is provided by the crystallins, which are the proteins making up the lens of the eye. In different types of vertebrate animal quite different sorts of protein have been brought into service to make a transparent lens and although they serve the same physiological function, they have virtually no sequence in common.

Implications of natural selection

Although the principle of natural selection seems very simple, the concept has some remarkably complex implications and these have been explored in great detail by theorists over the last 50 years.

Because it is capable of generating novelty, natural selection is in principle more interesting than neutral evolution. But there is still rather little information at the level of genes about what actually happened in evolution to generate the observed differences between species. The problem is that the mutations underlying a change may not be in those genes that normally control formation of that aspect of the organism. For example, the identity of the different segments in animals, in terms of their appearance and the types of appendages which they bear, are controlled by a family of genes called Hox genes. In experiments on insects it is possible to change the identity of segments in a predictable manner by introducing or removing individual Hox genes. However when two related species of insect are examined which differ in some aspect of segmental pattern, it is often found that they differ not in the presence of specific Hox genes, or even in their expression, but in other, unknown, genes, which modify the responses to the Hox genes. It seems that an intelligent designer, such as a modern developmental biologist, would do things differently from the way they actually happened in evolution.

Given that it is rather difficult to identify exactly what happened in evolution even to bring about those changes in structures whose formation is well understood by developmental biologists, it will probably be very difficult indeed to identify what happened to genes to generate more complex types of change, such as those concerned with animal behaviour. In fact, the 'genes' that are discussed in this context are mostly hypothetical, existing only within the discourse of evolutionary theory.

There are two sorts of hypothetical gene. One is the normal version which is responsible for a universal behavioural characteristic of a species. For example, male lions who take over a pride have a propensity to kill the existing cubs of other males. This aspect of lion-ness is said to be 'in the genes' although the actual genes that control this characteristic will probably always remain unknown. The other sort is a hypothetical gene variant: the new mutation which confers a new behaviour different from the current norm of the species. For example, suppose a mutation appeared which caused the carrier to sacrifice himself for others of the species. Could such a gene variant ever become established in the population, and if so, under what circumstances?

Problematic types of selection

If you look at an old-style palaeontology book it is likely to contain diagrams of the evolution of animal groups at the level of genera or families which change in evolutionary time. But once you start thinking about natural selection, you realise that whole groups of animals cannot evolve into others in this way. Families and genera are composed of species, so evolution for a higher group involves an independent parallel pathway of evolution for each of the component species, which seems unlikely. All new species must start as individual organisms which have acquired a reproductive advantage as a result of some mutation. They will pass the new variant to their offspring and because of the reproductive advantage, over many generations, the new variant will spread through the breeding population and become the normal form of the relevant gene. Once enough new variants have become established in this way the morphology of the creatures will have become changed enough for biologists to agree that it is a new species. If a new population becomes physically isolated from the old one, or if a mutation reduces mating opportunities between old and new, then the old species may persist along with the new one. Whether natural selection can operate at the level of higher groups has been very contentious. But it is important to recognize that there is an asymmetry in relation to higher group selection between extinction and progressive evolution. If the environment changes sharply this may bring about the extinction of a whole species or even a whole genus or family, and in this sense higher group selection does occur. But in the sense of development of evolutionary novelty, nothing can occur unless it is to the reproductive advantage of an individual carrying a new gene variant.

Since all selection is due to individual reproductive advantage, this raises problems for explaining the appearance and persistence of those aspects of organisms that seem to militate against individual reproductive success. One such area is sex. Sexual reproduction is almost ubiquitous among higher organisms, although a few asexual groups do exist. Since sex inevitably involves a fraction of the population (males) making only a small material contribution to the offspring, a mutation causing asexual behaviour should double the rate of reproduction and therefore be strongly selected for. But, since most organisms are sexual, this evidently does not happen very much. Why not?

Another area is altruistic behaviour. If one animal sacrifices itself to save others of its species, which certainly occurs in nature, it seems to be reducing its own reproductive chances and thereby allowing gene variants carried by the other individuals to prevail in the population. Explaining the origin and persistence of altruism has been a major theme in modern evolutionary theory.

These problems, and possible answers to them, were popularized in *The Selfish Gene* (1976) by Richard Dawkins, derived from ideas of William Hamilton (see later in the chapter). The selfish gene is not, in fact, a gene causing selfish behaviour. It is rather a way of looking at evolution that is gene-centred. Generally what is good for the organism is good for the propagation of the gene variants it carries. But sometimes, as for sex and altruism, maximizing inheritance of gene variants that bring them about seems at first sight to be of advantage to the group but of disadvantage for the individual, and hence it requires further explanation.

Sex

In mammals, sex is determined by the chromosome complement. Unlike all the other chromosomes, the sex chromosomes come in distinct X and Y varieties. Males are XY and females are XX. It might seem that this is a rigid mechanism that automatically generates equal numbers of two sexes, and which cannot be changed. However sex determination systems are actually very varied indicating that they have changed a lot in evolution. For example in birds the males have two similar sex chromosomes and the females have dissimilar ones.

There are many possible explanations for the persistence of sex, but they all tend to focus on the idea that sexual reproduction allows more flexibility in evolution because it recombines the gene variants from the two parents every generation. For an asexual creature to acquire two different specific gene variants requires a double mutation, which would be very unlikely to occur. But for a sexual population even two very rare variants will eventually be brought together by the independent assortment of chromosomes and by recombination within chromosomes. Thus sexual populations are more flexible in response to selection because they can generate novel combinations of variants more quickly. This is fine but it implies selection from an environment which is changing all the time, in every generation, because the selection has to be strong enough to resist the twofold advantage of reproduction rate immediately acquired by an asexual mutant. In general the environment does not seem to change that fast or that often.

One possible explanation that has been investigated to some extent is the so called 'Red Queen hypothesis'. It is named after the character in *Alice through the Looking Glass* who explained that in her country you always have to keep running faster just to stay where you are. This theory supposes that all organisms are infested with parasites. Genes in the host encode proteins that confer resistance to the parasites, and gene variants exist such that some variants confer greater resistance to some strains of parasite. The parasites carry genes that encode proteins enabling them to evade the resistance of the hosts. These also have gene variants that confer evasion to particular gene variants of the hosts. In this situation the parasitic infection produces continual new environments for the hosts because the parasites are selected to increase the frequency of gene variants that evade resistance most. Simultaneously the hosts are selected to promote the frequency of gene variants that impart resistance to the most common variants of the parasites. An asexual population of hosts will succumb to the parasites because it will only be able to produce new resistant gene variants by mutation, a much slower process than the selection of the parasite to evade resistance. A sexual population will be more successful because the process of sex continually brings together new combinations of the existing gene variants in the population and so continually produces organisms resistant to the most common forms of parasite.

This theory has been investigated in organisms such as snails and fish where sexual and asexual races coexist, and in many cases it is found that the asexual races do suffer higher levels of parasitism. In an experimental study, nematodes which may exist either as self-fertilized hermaphrodites or as sexual strains were infected with the same bacteria and maintained for many generations in the laboratory. In this artificial co-evolution situation, the self-fertilizing hermaphrodites go extinct after a few generations, whereas the sexual strain is able to maintain resistance long term.

So the Red Queen hypothesis has some evidence in its favour. It probably cannot explain the persistence of all sex everywhere because co-evolving parasites are not completely ubiquitous. However it serves as an example of how natural selection theory can explain certain aspects of evolution that seem at first sight to be incompatible with it.

Altruism

Animal behaviour is replete with examples of altruism. For example soldier termites engage in combat, sometimes to the death, to facilitate the flight of the reproductive termites. Dominant male baboons stand guard over foraging groups and will confront predators while the rest of the troupe retreats. Female nighthawks leave the nest when a predator approaches and settle in front of it feigning injury to distract attention from the young (Figure 19). These things have been known for a long time and in 1902 the Russian biologist Peter Kropotkin argued in his book *Mutual Aid* that cooperation and mutual aid are the most important factors in the evolution of species and their ability to survive. The zoologist V.C. Wynne Edwards proposed in 1962 that animals were able to limit their reproductive rate for the good

19. Nighthawk. Altruistic behaviour by the female nighthawk is seen when, to distract predators from the nest, she either droops her wings (left) or outstretches them (right)

of the species as a whole. This would occur typically when the population density was growing too high for the available food supply. He further proposed that mass swarming of animals served the function of allowing them to sense the population density. However, natural selection seems on the face of it to be incompatible with evolution of any type of altruistic behaviour. If such a behaviour existed and a mutation emerged which caused its carrier to cheat the system, for example by continuing to reproduce at a high rate while others reduced their reproduction, then this mutation would be favoured by selection and spread until it became the normal form of the gene.

The probable solution to the altruism problem can be traced back to J.B.S. Haldane, one of the founders of the modern synthesis of NeoDarwinism, who is reputed to have said: 'I would lay down my life for two brothers or eight cousins'. He identified that fact that a gene variant responsible for altruistic behaviour could propagate itself if it assisted the reproductive success of enough relatives who carried the same variant. Since brothers, on average, have 50 per cent of their variants the same, and cousins have 12.5 per cent the same, Haldane's calculation makes sense. In fact eight cousins only represent the break-even point and it would be more appropriate to lay down one's life for nine or more cousins because then you would be sure of giving your gene variants an advantage. The principle was developed mathematically by the theorist W.D. Hamilton who in 1964 introduced the concept of 'inclusive fitness'. Inclusive fitness is defined as the sum of all the fitness effects that the actor has on its relatives (including itself), each increment or decrement being weighted by the genetic relatedness to that recipient. Consequently, altruism may be favoured by natural selection despite the direct cost to the actor, so long as it provides a sufficiently large benefit to sufficiently related recipients.

So, for example, the soldier termites enable spread of their own gene variants by helping their reproductive siblings. The dominant male baboon is probably the father of many of the troupe he is aiding, each sharing 50 per cent of his gene variants. With sufficient ingenuity it turns out to be possible to explain just about any feature of animal behaviour by some type of argument based on inclusive fitness. For example instead of limiting their reproduction 'for the good of the species', birds will have an 'optimum clutch size' based on the maximum number of chicks that they can rear successfully given the prevailing conditions. Gene variants that predispose birds to pick up environmental signals about future food supply and feed this into the physiological controls of egg production would be favoured by selection. According to such arguments natural selection is compatible with exactly the same outcome as the discredited theory of Wynne Edwards.

Some very important examples of altruistic behaviour concern the social insects. Bees, wasps, ants, and termites include many species with highly differentiated social structures. These are varied and complex but in general there is a caste of reproductive individuals and a caste of non-reproductive workers. The workers do all sorts of chores to support the reproductives, and to look after their brood. Why would they do this? Darwin himself was aware that this could pose a problem for the theory of natural selection and suggested that selection was at the level of 'the family'. It turns out that the insect groups with this type of social structure all practice strict monogamy, or at least are thought to have done so when the social organization first evolved. Monogamy means that the inhabitants of one nest are all the offspring of a single queen and single male. This makes the relatedness high enough to explain why the sterile castes will work to support the success of the reproductive individuals. In some cases, such as the honeybee, the queen does mate with more than one male, but this is believed to be a newly evolved behaviour and to have emerged after the establishment of castes had become irreversible.

Inclusive fitness theory has explained many processes in behavioural ecology. Originally the gene variants were all hypothetical and the models were tested just by observing behaviour. But more recently some of the genes underlying behaviour and especially the colony structure of social insects, have been identified, so it is now possible to test models also by examining relevant gene variants.

Universal characters ascribed to 'genes'

'Sociobiology' is the trend in biology that examines animal behaviour from an evolutionary perspective. The term was popularized by the book of this name by E.O. Wilson published in 1975. It includes the considerations above relating to sex and altruism, and many other topics. It also includes the general idea that the normal behaviour of a species is 'in the genes'. In one sense it is an obvious truism, since behaviour like that shown by social insects is presumed to be hard wired into their nervous system and not the result of free and independent ethical decisions taken by each ant. However there has been a strong tendency to extend the arguments to human characteristics including perception, beliefs, customs, institutions, and so on. This obviously clashes with the traditions of social science that have sought explanations for human beliefs and behaviours in the dynamics and processes of human society and not in genetics at all. There has been, and will doubtless continue to be, considerable disagreement between scientists about the extent to which evolutionary models of animal behaviour can be extended to human beings.

The school known as 'evolutionary psychology' specifically aims to identify aspects of human behaviour that are the product of evolution. The founding manifesto is a book: *The Adapted Mind: Evolutionary Psychology and the Generation of Culture* edited by Jerome Barkow, Leda Cosmides, and John Tooby (1992). A prominent advocate for the school is Steven Pinker of Harvard University.

Evolutionary psychology differs somewhat from the aspects of evolutionary biology mentioned above in that it is not interested in genetic differences between people, only in the universal features of the human race. Indeed in the founding text it is baldly stated that there are no significant genetic differences between groups of humans, thus conveniently evading the contentious matters touched on in Chapter 5 above. The school dislikes accounting for human characteristics in terms of genetics versus environment and takes the view that all characters involve substantial interaction between the genetic inheritance and the environment, much of which consists of other human beings. At the same time, it does aim to expand the domain of 'nature' into the territory of the social sciences, particularly social psychology. For example it is recognized that characteristics such as the language spoken by an individual depends on the circumstances of his upbringing. But it maintains that the human mind has specialized modules for language acquisition which impose a necessary structure to the process, independent of experience, and indicate the existence of a genetically hard wired capability for learning language in a specific way. Similar arguments are deployed in relation to other behaviours such as mate preference, sexual jealousy, mother–infant emotional communication, and social contact algorithms.

Anthropologists usually argue against the existence of 'human nature' by pointing to variability between societies in more or less any human attribute. If you think of something that seems natural and universal, anthropologists will invariably find some tribe somewhere who do it differently. But evolutionary psychologists argue that the anthropologists are only looking at manifest properties and not at the underlying evolved system. They maintain that there are 'closed systems' equivalent to hard-wired behaviours, which are rather rare in humans, and 'open systems', like that responsible for language acquisition, where the potentiality is part of an evolved human nature but the manifest realization depends on culture.

A key feature of this tradition is the idea that very little human biological evolution can have occurred in the last 10,000 years, in other words during the Neolithic and historic periods. So our genetic heritage must date from the preceding Meso- and Palaeolithic periods when there was time for extensive evolutionary change, and which is documented to a limited degree by the human fossil record. In other words: 'modern humans have stone-age brains'.

The general case seems plausible but at the same time it is hard to identify real limits to human belief or behaviour. For example, some people choose not to have children, despite the likely biases in favour of reproductive behaviour arising from natural selection. Men may have some inherent aggressive tendencies but in modern societies these are firmly kept in check by social sanctions and by the legal system. The human mind may not be entirely a 'blank slate', but it does seem to be a slate on which the inscriptions can be changed.

Debates relating to sociobiology and to evolutionary psychology will doubtless go on for many years. From the point of view of the study of genes the important point about these disciplines is to realize that they involve a commitment to genetic determinism in general but they do not deal with any identified genes. It is always possible to think of a potential evolutionary explanation for any current trait, but proving that it actually happened is another matter. As we have seen this is even hard to do for those features of the anatomy whose developmental biology is well understood. In the end, we must accept that the gene concept is very broad and extends to areas where actual genes have not been, and may never be, identified.

Conclusion: the varied concepts of the gene

A principal object of this book has been to convey how many different concepts of 'the gene' there are, and how soundly they are based.

The gene of molecular biologists is a defined stretch of deoxyribonucleic acid (DNA) encoding a specific protein or ribonucleic acid (RNA). The structure, function, and regulation of many genes are understood in substantial detail. The genes of many different types of organism have been catalogued in the form of fully sequenced and annotated genomes. The same gene differs from one species to another by a number of neutral substitutions and these differences can be used for refining the classification of organisms.

Within a breeding population there is always some genetic variation. For example, different human individuals differ by about 0.1 per cent in genome sequence. There are some genes which, when mutated to inactivity or inappropriate activity, produce clear effects on the whole organism. However most variable traits arise from the action of many gene variants.

The 'genes' of human population studies are DNA markers differing between individuals. They are usually not in actual genes but in DNA outside genes. A specific subset of markers (certain simple tandem repeats) are used for forensic identification.

The 'genes' of quantitative genetics are mostly unknown variants at multiple loci which are presumed to have small additive effects on the character in question. The theory of quantitative genetics has been useful for informing programmes of animal and plant breeding and it is also used to calculate 'heritabilities' for human traits including some diseases.

The 'genes' of evolutionary studies were originally hypothetical gene variants, usually affecting animal behaviour in a certain way, postulated to exist for the purpose of creating models of evolutionary processes operating through natural selection. However some relevant gene variants can now be observed.

Finally there is the 'in our genes' type of gene where the bases of universal features of human behaviour or social organization are presumed to be determined by the genome in a similar way to anatomical or physiological characters, but no actual genes are examined.

At all these levels the gene concept has been important for society. Specific genes are the basis of much biomedical research and the biotechnology industry. The more hypothetical types of gene have been deployed to support ideas in the social and behavioural sciences about possible biological bases for all types of desirable or undesirable human traits.

In my view, the gene concept is at its most useful when considering single identified genes. The further away speculation runs from this core, the less reliable and more prone to misuse it becomes. However, the purpose of this book is not to tell readers what to think; it is rather intended to help them to evaluate the controversial issues on the basis of knowledge about real genes and their properties.

Further reading

There are numerous textbooks of genetics for use in higher education. These are just three of them:

J.D. Watson, T.A. Baker, S.P. Bell, A. Gann, M. Levine, and R. Losick (2014). *Molecular Biology of the Gene*, 7th edn. Upper Saddle River, NJ: Pearson. (Focus on molecular genetics.)

A.J.F. Griffiths, S.R. Wessler, S.B. Carroll, and J. Doebley (2012). *Introduction to Genetic Analysis*, 10th edn. New York: W.H. Freeman. (Wider coverage.)

L. Hartwell, L. Hood, M. Goldberg, A. Reynolds, L. Silver, and R. Veres (2010). *Genetics: From Genes to Genomes*, 3rd edn. New York: McGraw Hill. (More emphasis on genomics.)

The following are books written for the general public:

F.H. Portugal and J.S. Cohen (1977). *A Century of DNA: A History of the Discovery of the Structure and Function of the Genetic Substance*. Cambridge, MA: MIT Press. (Very informative about early work on DNA.)

J.D. Watson (1968). *The Double Helix*. London: Weidenfeld and Nicholson. (A racy account of the discovery of the 3D structure of DNA.)

A.M. Leroi (2003). *Mutants: On the Form, Varieties and Errors of the Human Body*. London: Harper Collins. (More readable than most books on human genetics.)

R.J. Herrnstein and C. Murray (1994). *The Bell Curve: Intelligence and Class Structure in American Life*. New York: Free Press.

(Very controversial book arguing for heritable differences in IQ between both individuals and groups.)

S. Rose, R.C. Lewontin, and L.J. Kamin (1984). *Not in our Genes: Biology, Ideology and Human Nature*. Harmondsworth: Penguin. (A fierce rebuttal of genetic determinism and sociobiology.)

R. Dawkins (1976). *The Selfish Gene*. Oxford: Oxford University Press. (A popular presentation of the gene-centred view of evolution pioneered by W.D. Hamilton.)

S. Pinker (2002). *The Blank Slate: The Modern Denial of Human Nature*. London: Allen Lane, Penguin Press. (A popular presentation of evolutionary psychology.)

The following are academic review articles which expand on topics mentioned in this book. The first two are of historic interest while the others give a more or less up to date picture.

A. Gulick (1938). What are the genes? I: The genetic and evolutionary picture. *The Quarterly Review of Biology*, 13: 1–18.

A. Gulick (1938). What are the Genes? II: The physico-chemical picture. *The Quarterly Review of Biology*, 13: 140–68.

A.O.M. Wilkie (1994). The molecular basis of genetic dominance. *Journal of Medical Genetics*, 31: 89–98.

M.L. Metzker (2010). Sequencing technologies—the next generation. *Nature Reviews Genetics*, 11: 31–46.

M.A. Jobling and P. Gill (2004). Encoded evidence: DNA in forensic analysis. *Nature Reviews Genetics*, 5: 739–52.

L. Kruglyak (2008). The road to genome-wide association studies. *Nature Reviews Genetics*, 9: 314–18.

P.M. Visscher, W.G. Hill, and N.R. Wray (2008). Heritability in the genomics era—concepts and misconceptions. *Nature Reviews Genetics*, 9: 255–66.

R.A. Kittles and K.M. Weiss (2003). Race, Ancestry, and Genes: Implications for defining disease risk. *Annual Review of Genomics and Human Genetics*, 4: 33–67.

D. Garrigan and M.F. Hammer (2006). Reconstructing human origins in the genomic era. *Nature Reviews Genetics*, 7: 669–80.

M. Burmeister, M.G. McInnis, and S. Zöllner (2008). Psychiatric genetics: progress amid controversy. *Nature Reviews Genetics*, 9: 527–40.

L.A. Dugatkin (2007). Inclusive fitness theory from Darwin to Hamilton. *Genetics*, 176: 1375–80.

G.E. Robinson, R.D. Fernald, and D.F. Clayton (2008). Genes and social behavior. *Science*, 322: 896–900.

Index